Integrated Power Management:
A Quick Start Guide

Published 2023 by River Publishers

River Publishers

Alsbjergvej 10, 9260 Gistrup, Denmark

www.riverpublishers.com

Distributed exclusively by Routledge

605 Third Avenue, New York, NY 10017, USA

4 Park Square, Milton Park, Abingdon, Oxon OX14 4RN

Integrated Power Management: A Quick Start Guide / by Vladimir Kopta.

Routledge is an imprint of the Taylor & Francis Group, an informa business

ISBN 978-87-7022-860-2 (paperback)

ISBN 978-10-0380-988-3 (online)

ISBN 978-1-032-62774-8 (ebook master)

A Publication in the River Publishers series
RAPIDS SERIES IN ELECTRONIC MATERIALS, CIRCUITS AND DEVICES

While every effort is made to provide dependable information, the publisher, authors, and editors cannot be held responsible for any errors or omissions.

Integrated Power Management: A Quick Start Guide

Vladimir Kopta

CSEM, Switzerland

NEW YORK AND LONDON

To my wife Bojana and my daughter Hana

Contents

Preface

This book is intended as a short introduction to the domain of integrated power management, primarily targeting low-power systems-on-chip. Such systems are present in a number of devices we use in our everyday lives, and power management is the core component needed to ensure a long battery life. The topic is essentially at an intersection of power electronics and analog IC design, built on well established and well known foundations, and applied to modern CMOS technologies.

It is primarily written for undergraduate and newly graduated students, as well as engineers that are getting familiar with the topic. In fact, my goal was to write a book that I would have wanted to read when first starting to work on power management. It covers some of the fundamental theoretical concepts, shows the essential ways to approach the design of complex blocks and provides an overview of some more advanced topics, potentially useful in the later stages of the design process. I have tried to avoid overly complex formulas as much as possible, and to put emphasis on the intuitive understanding of circuits and common trade-offs, which are, in my opinion, much more useful for practical design. Due to the limited space, some of the interesting topics that deserve a more thorough treatment are only briefly introduced. The curious reader is pointed to the additional literature that covers said topics, and is encouraged to dive deeper into the world of integrated power converters.

Belgrade, 2023   Vladimir Kopta

About the Authors

Vladimir Kopta received his B.Sc. degree in electrical and electronics engineering from the university of Belgrade, Serbia in 2011 and his M.Sc. and Ph.D. degrees from the Swiss Federal Institute of Technology in Lausanne (EPFL), Switzerland, in 2013 and 2018, respectively. Since 2018 he has been a research and development engineer at the Swiss Center for Electronics and Microtechnology (CSEM), where he has been involved in the design of low-power circuits and systems. The two focus areas of his work are RF circuits for wireless communications and power management for battery powered systems on a chip.

1

Introduction

Advances in CMOS technology and scaling of transistor feature sizes have enabled the integration of entire systems on a single silicon die, not bigger than a few square millimeters. Modern systems-on-chip (SoC) have reached considerable levels of complexity. They typically consist of one or more processors (CPU or a central processing unit) that interact with different peripheral modules, such as various hardware accelerators, digital interfaces, sensor interfaces or wireless radios. Regardless of the functionality of an SoC, each one requires a power management unit or a PMU. This is a sub-system, often integrated on the same die, that delivers power to each module of the SoC, aiming to satisfy different requirements set by these blocks while maximizing efficiency. The integrated PMU is the topic of this book. More precisely, we will deal with battery powered systems, such as the devices providing wireless connectivity and other functionalities, and targeting the internet of things (IoT) type of applications.

The SoCs considered here consume a relatively small amount of power. Relatively small means much lower power levels than what most power electronics engineers are used to, that is a peak power consumption in the range from a few milliwatts to few hundreds of milliwatts. Although the basic concepts and the underlying theory are exactly the same regardless of the power level, these low-power, miniature systems pose a number of specific challenges. The first one is, unsurprisingly, miniaturization in an attempt to integrate the PMU with the rest of the system and minimize the number of needed off-chip components and hence the cost of the device. Some of the design trade-offs and architectural choices are often driven by the miniaturization requirements and the general tendency to integrate as many components as possible. Battery powered devices actually spend most of the time in a sleep mode, where the consumption is minimized by preserving only the necessary functions. A good PMU must be able to provide high efficiency not only at peak power, when the full system is running, but also in sleep mode, as this is likely to determine the battery lifetime. The consumption in sleep mode may be well below 1 µW, and designing a PMU that will provide the needed efficiency in both cases is not an easy task. The design focus is not only on DC–DC converters and LDOs, but also on all the circuits surrounding them, as their performance and consumption may have a significant impact on the overall system performance.

The book aims to provide insights into all aspects of the low-power integrated PMU design, lay out a general design approach and point out the main trade-offs encountered in the described systems.

1.1 System Architecture

Modern SoCs consist of many different blocks, all of which may have different power requirements and impose different constraints. An example block diagram of an SoC is shown in Figure 1.1. Depending on the application, different analog and digital blocks are needed; some of the typically used blocks, such as the CPU, memories, accelerators, radios and analog to digital converters (ADC), are shown in the figure.

The reason a PMU is absolutely necessary is that there is a mismatch in the voltage provided by the battery and the voltage needed by the core circuits. The nominal supply of deep sub-micron CMOS technology nodes has reached levels below 1 V, while the standard battery voltage is often in the 3 V range. Furthermore, the battery voltage may vary depending on the charge state of the battery, and may drop considerably below the peak level over time. The internal resistance of the battery, that depends on the battery type and environmental factors, causes a dependency of the output voltage on the load current. The role of the PMU is to provide a clean and stable supply, that will enable proper operation of the core circuits in all conditions.

The PMU itself consists of several sub-blocks. Often the most important block is the switching DC–DC converter that converts the high battery voltage into a low voltage used by the circuits (step-down or buck converter). The efficiency of this block is crucial for the autonomy of the system, as it will practically carry the entire current of the SoC. Unfortunately, the switching

Figure 1.1: Example block diagram of a system-on-chip.

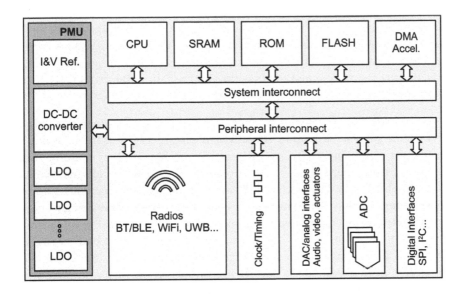

nature of the circuit causes a ripple in the output voltage. While some circuits, such as the digital CMOS circuits, are largely unaffected by this, others may be more sensitive. These are the circuits that operate with signals in the µV range, like the wireless receivers and sensor interfaces. In such cases a linear regulator is needed to suppress the ripple at the output of the DC–DC converter and provide a clean voltage. To maximize efficiency, the voltage difference at the input and output of the linear regulator (i.e. the dropout voltage) must be minimized. This is why they are usually referred to as the low-dropout (LDO) regulators.

The LDO regulator is a much simpler and smaller circuit than the switching converter, which is why it is possible to use one, or even multiple LDOs for each sub-block of the SoC. They can be used to provide different supply voltages to different blocks and to isolate sensitive blocks from noisy blocks. Note that the DC–DC converter is not the only switching block that may cause pollution of the supply, any digital circuit will do so as well. At the same time the LDOs act as supply switches and cut power to unused blocks to minimize leakage in the off state. This is why the general trend is to distribute the LDOs all over the chip and to separate supplies of different blocks.

The final part of the PMU consists of all the remaining necessary circuits that support the switching converter and the LDOs. By this we primarily refer to voltage and current references that define the output voltages, the clock generator for the switching converter, as well as all the control signal generators, glue logic, comparators, level-shifters and other interface circuits. All these auxiliary (or service) blocks are used to define the power-on sequence of the PMU, to start the system and to allow the CPU to talk to the PMU and control it through software.

Aside from the mentioned components, some SoC incorporate means to harvest energy from available sources and charge the battery in order to increase the autonomy. There are different energy sources that could be used: some that gather energy from the environment like the solar cells, mechanical or thermoelectric harvesters, or dedicated power delivery systems like the wireless chargers that transfer energy via electromagnetic waves. Since energy harvesters typically output a very low voltage, a step-up switching converter is normally employed to convert it into a high voltage used by the battery charger or the step-down converter to supply the core blocks directly. Such a step-up converter is used to extend the battery life through energy harvesting, in some cases even to allow the device to operate indefinitely, and is often part of the PMU.

1.2 Before We Begin

The following chapters deal with analysis and design of each of the parts of a typical PMU: linear regulators, switching converters and some of the commonly used auxiliary blocks. However, each of those blocks consists of one or more basic building blocks, such as switches, current mirrors, differential amplifiers, operational transconductance amplifiers (OTA), comparators, etc. Potential implementations of these basic blocks are shown and briefly discussed in some cases, but without thorough analysis and design guidelines. The assumption is that the reader is already familiar with these circuits. If this is not the case, there is an abundance of literature covering these topics, the two references that serve as a good starting point for any analog designer are [1] and [2].

This book deals exclusively with the integrated PMU, by this we mean integrated with all the other components in a low-power SoC in a standard CMOS technology. The devices that we have available to implement the PMU are MOS transistors, which often come in several different variants that support different voltages. Core MOS devices in today's processes often cannot survive more than 1 V from drain to source or from gate to drain. They are optimized for high speed digital circuits and are often a bit too fragile for use in a PMU. Fortunately, thick-gate devices are always provided alongside standard core devices. Thick-gate MOS transistors are used to drive off-chip circuits, and generally provide interface to the outside world. They have larger feature sizes and tolerate higher voltages, often close to the battery voltage, making them useful for PMU implementation. Typically, these are the devices that would be used as the main switches of a DC–DC converter. Other options, such as the BJTs or dedicated power transistors are not considered here (aside from the parasitic BJTs used in voltage references), as they are not needed in this application.

It is worth mentioning that all the derivations in this book use the EKV MOS transistor model [3]. This is to avoid any confusion that might arise from the equations that are perhaps slightly different than those in standard analog textbooks. In most cases there is no difference, for example small signal schematics are the same regardless of the underlying model, but the expressions for the MOS transconductance G_m and the output conductance G_{ds} may be different. The advantage of the EKV model is the continuous expression valid from weak to strong inversion, that is particularly convenient for low-power design. The few equations worth mentioning here, in the PMU context, are the approximate expressions for the drain current in strong inversion:

$$\text{Saturation}: I_D = \frac{\beta}{2n}(V_G - V_{T0} - nV_S)^2 \tag{1.1}$$

$$\text{Linear}: I_D = \beta(V_G - V_{T0} - \frac{n}{2}(V_D + V_S))(V_D - V_S). \tag{1.2}$$

The parameter $\beta = \mu C_{ox} W/L$ is determined by the transistor geometry, mobility and the oxide (dielectric) capacitance and n is a constant slope factor. All the voltages are referred to the bulk terminal. In the cases where the source is tied to the bulk, $V_S = 0$ simplifying the equations a bit. In weak inversion the drain current is given by

$$I_D = 2n\beta U_T^2 e^{\frac{V_G - V_{T0}}{nU_T}}\left[e^{-\frac{V_S}{U_T}} - e^{-\frac{V_D}{U_T}}\right]. \tag{1.3}$$

The transistor is considered in saturation when $V_D > 4U_T$, and the last exponential term can be dropped out of the equation. The shown expressions can be used for basic calculations, to derive G_m, or the on resistance of a MOS switch, which will be useful in the chapters that follow.

1.3 Outline

The following chapters detail different sub-blocks of a typical integrated PMU:

- *Linear regulators*: Fundamental aspects and specifications of an LDO are first discussed, followed by a description of different parts of an LDO, the common design approach and the insight into various trade-offs. An overview of practical LDO techniques from the literature is provided.
- *Switching regulators*: Inductive and capacitive switching converters are described. Fundamental theoretical concepts are revised, complemented by the efficiency optimization analysis and design guidelines. Common control loops are described, with details of transistor-level implementation and finally, an overview of useful low-power techniques is given.
- *Auxiliary circuits*: The chapter focuses on all the service blocks required by a typical PMU, detailing the interfaces and the example start-up sequence. Practical examples of voltage and current references, clock generators, comparators and interface circuits are given here.

2

Linear Regulators

Linear regulators are circuits that take a high, unregulated and potentially noisy voltage as input, and provide a clean, stable voltage at a correct level to their load circuit. They can be used to convert the battery voltage into a low voltage used by the CMOS circuits. The problem with this approach is that the power loss in a linear regulator is directly proportional to the dropout voltage (i.e. the difference in the input and the output voltage), which may prove to be considerable in the mentioned example. To avoid wasting power, which is quite precious in battery powered devices, it is better to use a high-efficiency switching converter to lower the battery voltage to a level close to the core voltage with minimum losses. A linear regulator is then used to cover the remaining difference without a significant penalty in the overall efficiency. Because linear regulators are almost exclusively used when the dropout voltage is low, they are usually referred to as the low dropout regulator or the LDO, and the two terms have practically become interchangeable today.

In this chapter we will look into LDO design, starting with some generalities and listing all the important attributes of an LDO. Then we will go through different parts of the LDO, roughly following the order in which you could approach the design. Unfortunately, there is no magical, one-size-fits-all type of solution, instead we will try to explain different trade-offs and cases you might encounter in practice and point the reader towards a possible solution.

Figure 2.1: LDO with a common source output.

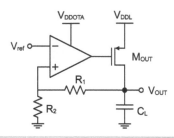

2.1 LDO Fundamentals

As briefly explained, the LDO provides a clean, stable supply and isolates its load from the noisy environment. The very basic schematic is given in Figure 2.1. It consists of the output, or pass transistor and the error amplifier (EA), usually an operational transconductance amplifier (OTA) or an operational amplifier (OA). The output transistor carries the load current, while the EA assures that the output voltage is at a desired level through negative feedback. In the LDO from the figure the output voltage is set by the resistive divider and the reference voltage, such that $V_{out} = (1 + R_1/R_2)V_{ref}$. A unit feedback could be used instead, in which case $V_{out} = V_{ref}$.

The first thing that comes to mind when talking about LDOs, is the regulation of the output voltage. The output voltage will vary with the load current of the LDO, this is known as load regulation, or with the LDO supply voltage (V_{DDL} from Figure 2.1), which is called the line regulation. Typical curves are shown in Figure 2.2. Due to the finite gain of the EA, and finite loop gain, the output voltage changes with the load current. Typically it decreases slowly with the load current until it reaches a limit (a sort of knee region) after which the output quickly drops to 0. The slope of the curve depends on the loop gain, output range of the EA, pass transistor geometry and the dropout voltage. Higher loop gain keeps the voltage more constant, however it doesn't extend the current limit significantly. In the common-source configuration of the pass transistor the limit might come from the EA output voltage range (i.e. the gate voltage of the pass transistor). It is generally a good idea to use an EA that can reach a near-zero voltage together with a PMOS pass transistor. Otherwise, increasing the output transistor width or the dropout voltage could help with the current capacity of an LDO. Essentially, by doing this you keep the output transistor away from the linear region, which helps with both line and load regulation. In modern SoCs the LDO supply will often come from a

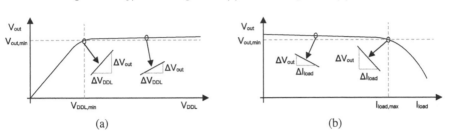

Figure 2.2: Typical line regulation (a) and load regulation (b) curves.

DCDC converter, whose output is also regulated, so the static line regulation is somewhat less of a concern. What is of concern is the transient response of the LDO and the power supply rejection (ratio) or PSRR.

This is probably the second most important thing to consider regarding LDOs, as they must provide a noise free supply to their load. In general, a common approach with modern SoCs is to have different LDOs supplying different blocks of the system. The idea is to isolate sensitive analog blocks such as sensor interfaces, LNAs, ADCs and others from noisy blocks such as the CPU, DCDC converter or the ADC. Note that the ADC is both a sensitive block (due to high resolution and a potentially small input signal) and a block that can generate a lot of supply noise due to switching activity. For a small disturbance, the amount of isolation an LDO provides is quantified by the PSRR. More details will be discussed in the following text, but for now, perhaps a few intuitive comments are in order. A low-frequency PSRR should correspond to the derivative of the line regulation curve. For a high dropout voltage it should be fairly good and degrade as the input approaches $V_{DDL,min}$. The frequency characteristic of the PSRR depends on several factors. Inside the bandwidth of the error amplifier negative feedback regulates the output and suppresses any disturbances coming from the input. Unfortunately, the EA bandwidth is limited and at higher frequencies the PSRR will be a function of the pass transistor characteristics, such as the output conductance and parasitic capacitances, as well as the load capacitor of the LDO and the load properties. In principle, a large load capacitor should reduce the load impedance, and improve the PSRR. There are issues with this approach; for one, a large capacitor requires a large silicon area, which is expensive. It is possible to use an off-chip capacitor, but that also increases the price and the size of the whole system. Furthermore, large capacitors could affect the stability of the LDO.

This brings us to the next point, the LDO stability. The problem is exactly the same as the one you may have encountered with common feedback amplifiers.

The issue here is that the LDO doesn't necessarily have a fixed or known load current, which affects the loop gain, and hence stability. In a general case it is necessary to assure stability for a large range of currents, going from almost 0 current all the way to several milliamps or even hundreds of milliamps, which can be a tedious task. The design problem simplifies quite a bit if a known load that draws constant current is supplied. The techniques for compensating an LDO are the same ones used to compensate any amplifier, the difficulty lies in the fact that you need to cover a range of load currents as opposed to a single value.

Another small signal parameter of the LDO is the output resistance, or impedance to be more precise. It quantifies how much changes in the load current couple to the output voltage. Ideally, the output impedance should be very small. Similar to the PSRR, the low frequency output resistance is the derivative of the load regulation curve. The closer you are to the current limit of the LDO, the worse it becomes. Inside the EA bandwidth negative feedback maintains the low output impedance, whereas outside it depends on the load capacitor and pass transistor characteristics. Techniques to improve output impedance are generally very similar to the techniques used to improve the PSRR.

So far we've mentioned small signal effects. Indeed, the PSRR and the output impedance are mere approximations that correctly model what happens for small signals. But, suppose you have a step in the load current that goes from 0 to I_{max} in a very short time interval. In this case the nonlinear effects come into play and small signal parameters will no longer accurately describe what happens. Typically, a step in supply voltage, or load current, results in a spike at the LDO output. Depending on the direction of the spike, we will talk about overshoot or undershoot of the output voltage. As a circuit designer, your task is to assure that under worst case conditions the output voltage does not exceed the specified limits. In case of an undershoot, there is a risk, for example, that some flip-flops in the CPU reset due to the low supply voltage, which then cause erroneous behavior. The extreme limit for overshoot is the technology supply voltage limit, that you should not exceed in order to avoid degradation of transistor performances and reduction of product lifetime. In a less extreme case, you will want to avoid inducing nonlinearities in the circuit behavior due to supply voltage variations. One example could be the resistance modulation of the switch in the input S&H circuit of an ADC. The amplitude of the voltage spikes depends on several parameters. First of all, it depends on the amplitude and rise or fall time of the voltage or current step. As this can make a real difference, it is useful to know your load and to verify realistic scenarios to avoid over-engineering the LDO. Although this is a safer option, it is better to save time and not tackle the problems that need not be solved.

If overshoot and undershoot pose a real threat, then it's worth revising the design and finding ways to improve it. In general, the same techniques that improve the PSRR and output impedance help with transient behavior. To some extent the spikes can be reduced by increasing the EA bandwidth, but bear in mind that stability requirements may limit the achievable bandwidth. It is also worth looking into the slew rate of the EA as it might limit the reaction time of the LDO. Improving the slew rate or bandwidth practically always leads to the increased consumption of the EA, which illustrates one of the basic trade-offs. Still, the feedback loop will never be able to react fast enough to counter very steep steps in the load current. The only cure for such cases is a larger decoupling capacitor. More charge stored on a larger capacitor results in a smaller output voltage variation, which is paid in the silicon surface.

One other important case of transient behavior that deserves our attention is the start-up of the LDO. Battery powered SoCs never operate continuously, instead parts of it will be turned on and off as needed in order to maximize autonomy. In order to avoid wasting too much time in transition, it is usually good to have the LDOs turn on as fast as possible. This is where that big capacitor, that seems to solve a lot of issues, becomes problematic. In order to quickly charge this capacitor, a high current pulse (inrush current) would be necessary. However, most batteries have a limit in the current they can provide before their capacity is affected. For this reason, in such types of LDOs a current limiter may be needed to prevent the high inrush current. This feature is called a soft start. Clearly, the more you limit the current the more time it will take to start the LDO, which is a clear incentive to use the minimum size of the load capacitor.

As with all analog blocks, one must take care of noise when designing an LDO. Depending on what the load is, it may or may not be the limiting factor. One simple example is the DCO, where the LDO noise might couple from the supply to the DCO output and increase phase noise. LDO noise is typically quantified by the total integrated noise that translates into the equivalent noise voltage, although in some examples the shape of the noise spectrum is also relevant. It is important to note that the noise at the LDO output depends not only on the LDO noise, but also on the noise of the voltage reference. From the perspective of the reference, the LDO acts as a buffer, meaning that practically all of the reference noise appears directly at the output, possibly with some gain if a resistive divider is used. This is why it is important to verify the two circuits together. It is easy to overlook this fact and simulate the LDO with an ideal voltage source as a reference, which may hide some of the effects. Reference noise is commonly filtered using an RC filter with a high series resistance, a simple solution, but it is important to remember that the time constant of the filter may impact the overall start-up time of the LDO.

Output voltage variation is another important parameter of every LDO. Due to random mismatch, some offset will always be present at the EA output, which will then produce deviation of the output voltage from the desired value. The dominant source is typically the offset of the differential pair or the current mirror in the first stage of the OTA. A common way to approach this problem is to increase the size of the differential pair as mismatch is inversely proportional to the square root of the transistor area. At the same time, increasing the transistor width, while keeping the same bias current, reduces the inversion coefficient, which is another way to improve matching of the differential pair. For current mirrors, the case is slightly different as the matching is better in strong inversion, meaning that it's beneficial to increase transistor length [3]. Off course, depending on the supply voltage and the available headroom it may not always be possible to bias the current mirrors in strong inversion (this is especially the case in deep sub-micron technologies). In the end it is up to the designer to prioritize and decide what needs to be sacrificed. Aside from analog techniques, it is also possible to improve offset by implementing digital calibration of the EA, although this is rarely needed for the LDOs. Note that if the resistive divider is used, the total offset at the LDO output will be the EA offset multiplied by the same factor as the reference voltage. It is important to remember that the reference voltage variation also contributes to the output voltage variation and needs to be taken into account. Sometimes the reference can be trimmed after production, but this will add to the cost of the chip. Finally, it is important to account for the voltage shift when considering the overshoot and undershoot, as this will eat up part of the headroom.

The power losses in the LDO come from the EA, the resistive divider and from the dropout voltage. The challenge in LDO design is to minimize losses, while maintaining all the so far mentioned parameters within the defined limits. A typical trade-off for the EA is between consumption and bandwidth (speed). The bandwidth directly affects the PSRR, noise and transient behavior, so these specifications will implicitly dictate the minimum consumption of the EA. More often though, the efficiency will be limited by the dropout voltage. Assume a 0.1 V dropout voltage for a 0.9 V supply (a rough number for a 22 nm node for example). If the EA consumption is negligible compared to the load current, 10% of the power will be dissipated in the pass transistor. It is clear that the dropout voltage should be minimized, however low dropout voltage will degrade all of the LDO performance parameters. Finding the sweet spot, and a good balance among all the requirements is the true challenge in LDO design.

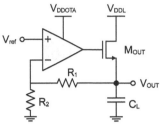

Figure 2.3: LDO with an output source-follower.

2.2 Output Transistor

When starting the design of an LDO, the first thing to do is to choose the output transistor. Normally this shouldn't be a tough one as the choice is limited to a PMOS (common-source), seen in Figure 2.1, or an NMOS (common-drain or the source follower). Moreover, the choice is likely to be dictated by the overall system architecture and the available supply voltages.

An LDO with a source follower output is shown in Figure 2.3. It should be immediately apparent that there is a small problem here. We said already that we wanted to minimize the dropout voltage, a reasonable expectation would be the range from 0.1 V to 0.2 V. The V_{GS} of the output transistor is set by the negative feedback in order to support the load current, and is normally above the V_{TH}. Even in very advanced technology nodes the threshold voltage will rarely be lower than 0.4 V, and if the output transistor is a thick-gate transistor the threshold voltage will be closer to 0.7 V. The conclusion is that the EA supply must be higher than V_{DDL} in order to drive the output transistor properly, otherwise the large dropout voltage would make this design highly impractical.

Fortunately, the EA shouldn't consume a lot, which reduces the burden on the circuit supplying the higher voltage. There are implementations that include a simple charge pump voltage doubler with the LDO [4] and use it to supply the EA. If you have a system with multiple LDOs, a single charge pump may be sufficient to provide a high supply for all of them. Otherwise, it might be the case that there is already a sufficiently high battery voltage in the system, which you might use without regulation to supply the EA. As long as the the total consumption of all error amplifiers is negligible to the consumption of the system, the overall efficiency shouldn't be impacted. A DC–DC converter will commonly be used to efficiently convert the battery voltage

Figure 2.4: Example schematic with the OTA supplied from the battery and the output source follower from a DC–DC converter.

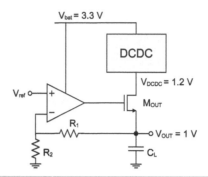

down to V_{DDL}, which is where the high load current is flowing. The case is illustrated in Figure 2.4.

It is interesting to look at the two topologies and to compare them. Clearly, the common-source approach is more general and can be used regardless of the overall system architecture. The question is: are there any benefits to the source-follower pass transistor that would justify the added complexity? To answer this question it is good to first look at the loop gain. For the CS topology, the loop gain is given by

$$A_{LG,CS}(s) = A(s)G_m \frac{R_L R_2}{R_L + R_1 + R_2}, \tag{2.1}$$

where G_m is the transconductance of the output transistor. The output resistance of the transistor is assumed much higher than parallel combination of R_L and $R_1 + R_2$, but this may not always be the case. Note that the error amplifier gain $A(s)$ is frequency dependent. Typically this will be a first order low-pass function $A(s) = A_0/(1 + s/\omega_p)$, for a single stage error amplifier, and it can be more complex if multiple stages are used. In the case of the output SF the loop gain is given by

$$A_{LG,SF}(s) = A(s) \frac{R_2}{R_1 + R_2} \frac{G_m(R_L \parallel (R_1 + R_2))}{G_m(R_L \parallel (R_1 + R_2)) + 1} \approx A(s) \frac{R_2}{R_1 + R_2}. \tag{2.2}$$

We can make several conclusions based on the above two equations. The CS topology should typically provide higher loop gain, as the output transistor itself contributes some voltage gain. It should therefore provide better voltage regulation at its output. This is only true if the load resistance is sufficiently high, otherwise, if R_L is comparable to G_m there may not be a significant benefit to the CS topology. A useful thing to mention is that the SF is less sensitive to the load, a somewhat intuitive result given its lower output resistance compared to the CS.

If we calculate the output resistance for the CS we get

$$Z_{out,CS}(s) = \frac{R_1 + R_2}{A(s)G_m R_2} \parallel (R_1 + R_2) \parallel \frac{1}{G_{ds}} \approx \frac{R_1 + R_2}{A(s)G_m R_2}, \qquad (2.3)$$

while for the SF case we get

$$Z_{out,SF}(s) = \frac{1}{G_m(1 + A(s)\frac{R_2}{R_1+R_2})} \parallel (R_1 + R_2) \parallel \frac{1}{G_{ds}} \approx \frac{R_1 + R_2}{A(s)G_m R_2}. \qquad (2.4)$$

Assuming sufficiently high gain of the error amplifier, the output resistance is roughly the same in both cases. Even though the loop gain is higher in the CS topology, the lower output impedance of the SF compensates for this and in the end yields the same result, giving no particular advantage to either topology.

The next parameter to compare is the PSRR of the two topologies. For this calculation we are looking into the output response due to perturbations in V_{DDL}, assuming there are no disturbances in V_{DDOTA}. This may or may not be the case in reality, but the PSRR is practically always dominantly determined by the input to output coupling through the pass transistor (and the calculation is slightly easier). Under the given assumptions, the PSRR of the CS LDO is given by

$$\frac{V_{out,CS}}{V_{DDL}} = \frac{G_m + G_{ds}}{G_{ds} + \frac{1}{R_L} + \frac{1}{R_1+R_2} + G_m \frac{A(s)R_2}{R_1+R_2} + sC_L} \approx \frac{R_1 + R_2}{A(s)R_2}. \qquad (2.5)$$

The approximation is valid in the region where the EA gain is sufficiently high. Essentially, any perturbation on V_{DDL} is suppressed by the gain of the error amplifier. At frequencies above the amplifier cut-off frequency, the gain starts to drop and PSRR slowly degrades. For comparison, the PSRR of the SF LDO is given by

$$\frac{V_{out,SF}}{V_{DDL}} = \frac{G_{ds}}{G_{ds} + \frac{1}{R_L} + \frac{1}{R_1+R_2} + G_m(1 + \frac{A(s)R_2}{R_1+R_2}) + sC_L} \approx \frac{1}{A(s)\frac{G_m}{G_{ds}}\frac{R_2}{R_1+R_2}}. \qquad (2.6)$$

Again, owing to the negative feedback, the EA suppresses any signal coming from the supply, but here we have an additional term G_m/G_{ds} (intrinsic gain of M_{out}) that provides a superior PSRR compared to the CS topology. This is one of the main advantages of the SF output transistor. The result should be rather intuitive as in the CS case the perturbations occur at the source of the output transistor which directly affects its current. Neglecting the channel-length modulation, the same perturbations applied at the the drain of the SF have no impact on the output voltage.

Another benefit of the SF topology is that it offers higher compensated bandwidth. To understand this, we should have a look at the pole located at the LDO output in the open loop configuration. The output resistance of the CS transistor is high and equal to $1/G_{ds}$, and the loading capacitor will see this resistance in parallel with R_L and $R_1 + R_2$. In most practical cases R_1 and R_2 will take high values in order to minimize the added current, or might not even be there (in a unity buffer LDO configuration). Load resistor R_L could also be relatively high, even for a high load current (consider an analog block with a fixed bias current), which means the pole will be largely determined by the G_{ds} of M_{out}. On the other hand, this pole will be much higher for the SF topology, and approximately equal to G_m/C_L. If the dominant pole is determined by the EA, this will allow higher gain-bandwidth (GBW) for the same phase margin, and hence better dynamic behavior.

A downside of the SF LDO is the leakage in the off state, which will typically be an order of magnitude higher compared to the CS LDO. If leakage current is important, and it might be when the chip is in low power mode and the supply is kept on to preserve the value of some registers or memory blocks (this would be a retention or sleep mode, where you want to keep the processor context in order to restart the system quickly), but the rest of the system is off. There are a few options to improve leakage. You could push the pass transistor gate voltage below 0 using a charge pump, which only makes sense if the charge pump itself consumes next to nothing. Otherwise, a PMOS switch could be placed in series with the NMOS, but this increases the area.

One more practicality should be considered when discussing the output transistor. The case considered here, which is an SoC with integrated power management, imposes some additional constraints on the transistor types that can be used. If the core of the SoC is supplied by the technology nominal voltage, the pass transistor supply V_{DDL} must be higher. For this reason you might need to use a thick-gate transistor at the output[1]. In principle, if the difference $V_{DDL} - V_{OUT}$ is always lower than the technology limit, a core output transistor might be an option. This is very tempting as the core transistors are smaller, provide better transconductance, and generally lead to better PSRR and bandwidth, but require special care. First you have to make sure that none of the voltage limits are breached in the steady state. For example, for the common-source topology you should make sure that the V_{SG} is sufficiently low. This means you may need to limit the output voltage of the EA to prevent it from pulling the gate too low at high load current. In addition, you must make sure

---

[1]In every technology node you will find core transistors and thick-gate transistors that are able to support higher voltages. These transistors are needed to interface the outside world as other chips in the system might require higher voltage and logic levels. Commonly, outside the chip you will use 1.8 V or 3.3 V, while the core of the deep sub-micron technology nodes is supplied by less than 1 V.

that no voltage exceeds the limit during start-up and shut-down of the LDO and in general when the LDO is turned off and V_{DDL} is on. You may have to add a thick-gate switch to avoid reliability issues, in which case you might end up with a similar total size of the output transistor (that is a combination of the switch and the regulating output transistor), as if you had simply chosen a thick-gate pass transistor.

2.3 Error Amplifier

The error amplifier is an OTA that drives the output transistor. There is already quite an exhaustive literature on OTAs, and here the focus is on aspects related to LDOs. Contrary to generic OTAs, the input voltage range will either be rather small (if you want to have the option to tune the output voltage), or fixed. The input range might force you to choose a PMOS or an NMOS input differential pair, depending on the reference voltage and the available supply. At the same time, the output voltage range will be largely determined by the choice of the output transistor. It is important to make sure that the OTA can provide $V_{out} + V_{GS}$ for the NMOS output or $V_{DDL} - V_{GS}$ for the PMOS output, where V_{GS} depends on the output current and the transistor size. Make sure that for a given load current the output transistor is sized such that there is enough voltage headroom in all PVT corners, accounting for some margin for the OTA. For example, ensure that $V_{DDOTA} - V_{ds,sat} > V_{out} + V_{GS,out}$ in the worst case to make sure the OTA output is always in saturation. Fortunately, you will practically never have a case where you need to provide a true rail-to-rail output voltage, which slightly simplifies the design.

When starting with the OTA design, good advice is to start with a simple topology. The simplest OTA is the common differential pair, as shown in Figure 2.5(a), together with an NMOS output stage. In terms of output voltage range it seems to fit the purpose very well. If you were to use the same OTA to drive an output PMOS you could encounter some difficulties at supplies below 1 V. If this is the case, one option would be to use a symmetrical OTA, as shown in Figure 2.5(b), which can achieve lower output voltage. It is now interesting to discuss the DC gain, GBW and consumption of the error amplifier. For the simple differential pair the voltage gain is $G_{m1}/(G_{ds2} + G_{ds3})$. Assuming the input differential pair is biased in weak inversion, G_{m1} will be proportional to the tail current of the differential pair. The same will be true for the output conductance of M_2 and M_3, meaning that increasing the bias current doesn't increase gain. Put more optimistically, you don't need high current to achieve high gain, a useful note to remember for low-power design. The bias current does, however, impact the gain-bandwidth product which is equal to G_{m1}/C_c.

Figure 2.5: Simple differential pair as an OTA (a) and a symmetrical OTA (b).

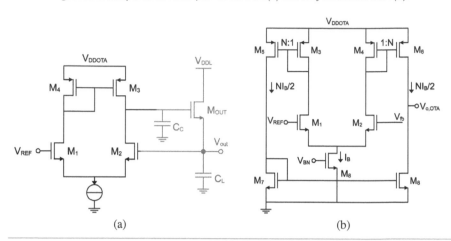

(a)　　　　　　　　　　　　　　　　　　　(b)

For the purpose of this discussion, assume that C_c is a combination of all the loading capacitors at the OTA output; if the compensation capacitance is not present it will be dominated by the gate capacitance of the pass transistor. For a fixed OTA load capacitance, increasing GBW requires an increase in the bias current. In general, when designing an LDO, the GBW of the first stage will also be limited by the frequency of the non-dominant pole, so an additional limiting factor comes from the stability requirement. In the symmetrical OTA an additional factor N is present due to the mirroring ratio, so the gain is equal to $NG_{m1}/(G_{ds2} + G_{ds3})$ and GBW to NG_{m1}/C_c. This factor comes with the additional current NI_B flowing in the two added current branches.

If higher DC gain is needed from the EA, a common approach is to use a cascode OTA. Figure 2.6 shows two implementations of a folded cascode OTA. A standard cascode OTA (telescopic amplifier from [1]), is often impossible to implement at supply voltages below 1 V, but may be a viable option if the OTA is supplied directly from a battery. The version from Figure 2.6(a) uses a "poor man's cascode," a convenient topology that doesn't require additional circuitry for biasing the cascode transistors. When sizing transistors $M_{3,4}$ and $M_{5,6}$, it is important to make $M_{3,5}$ sufficiently wider than $M_{4,6}$ in order to prevent desaturation of transistors. Note also that the bottom current mirror is not cascoded, here the output impedance can be increased by simply increasing transistor length. By using a folded cascode, additional current branches are introduced, this illustrates an example where low voltage imposes increased consumption due to limited headroom. The gain of the OTA in Figure 2.6(a) is $G_{m1}/(\frac{G_{ds2}G_{ds5}}{G_{m5}} + \frac{G_{ds8}G_{ds10}}{G_{m10}})$. Using a cascode stage essentially boosts the output

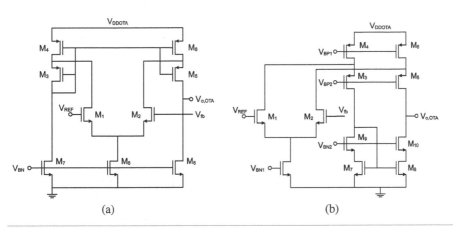

Figure 2.6: Simple differential pair as an OTA (a) and a symmetrical OTA (b).

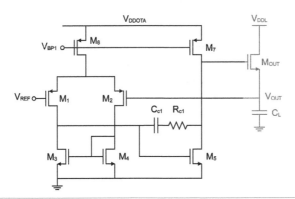

Figure 2.7: Two-stage OTA with an NMOS pass transistor.

resistance of the amplifier by a factor equal to the intrinsic gain of the cascode, but it doesn't affect bandwidth. For example, by using a cascoded OTA, you will improve the PSRR at low frequencies, which could be of importance for a sensor that needs to detect a very small, low-frequency voltage. At slightly higher frequencies the PSRR will remain unaffected. Higher bandwidth is necessary here, ultimately leading to higher consumption.

In some cases multiple stages may be needed. More often this will be the case with the NMOS output transistor as it doesn't contribute to loop gain. You may add a second stage to extend the output voltage range of the cascoded first stage. Another incentive to use a multiple-stage OTA is to improve the slew-rate, as single-stage OTAs are limited by the tail current source and the loading capacitance coming from the output transistor gate, which might prove to be fairly large. In the example from Figure 2.7 the NMOS in the second stage can provide large current to discharge the gate of M_{out}, however, the charging speed is still limited by the bias current of M_7. If the charging speed is more important a PMOS stage could be an option, otherwise the only way to eliminate slew rate is to use a class-AB error amplifier.

The difficulty of using multiple stages lies primarily in stability. Using a two-stage OTA means that you now have a three-stage amplifier to stabilize. To add to this, you need to make sure that it is stable for a wide range of output currents. Consider again the LDO from Figure 2.7. Owing to the bootstrapping action of the source-follower, the second stage doesn't see the full gate capacitance. For example if the SF gain is 0.8, only 20% of the gate capacitance will be seen (intuitively, if the gate and source voltages are almost equal, there will only be a small current going into C_{GS}, thereby lowering the effective capacitance). As a result the pole from the second stage might be at a fairly high frequency. It might then be sufficient to use Miller compensation across the second stage to stabilize the LDO. Note that in this case you could also connect the compensation capacitance to the output node to avoid loading the output of the second stage. In a general case, using multiple-stage EA will require special care for stability. A good place to start is [1], which provides various compensation techniques for multiple stage amplifiers.

2.4 Stability, Load and PSRR

It should be intuitively clear by now that the stability and PSRR are somehow related, and, as shown in Equations 2.1–2.6, are dependent on the EA bandwidth and the load of the LDO. Very often in the literature the load of an LDO or a DC–DC converter is modeled by a resistor. This implies that a high load current is equivalent to low output resistance, which may be nice as it improves the PSRR and pushes the output pole to higher frequencies, but may not correspond to reality. Furthermore, it could lead to erroneous conclusions and overly optimistic results. A typical example of an LDO load is an analog circuit biased by a constant current. The bias current, and hence the consumption of the load should ideally be independent of the supply, which means that you could have a high load current and a very high equivalent resistance – the worst case for an LDO designer. Another typical example is the digital circuits. Recalling the formula for dynamic power dissipation of CMOS gates, we could deduce that the average current drawn from the supply is proportional to $f_{CK} C_{eq} V_{DD}$, where f_{CK} is the clock frequency and C_{eq} is the equivalent capacitance of the entire digital block that also includes the average activity of the gates. Below the clock frequency a digital block will behave as an equivalent resistor with $R = 1/(f_{CK}C_{eq})$, but you shouldn't forget that a digital load is actually a train of short pulses with a relatively high amplitude and a significant high-frequency content. In fact, the reason why a separate digital LDO is commonly used in SoCs is to isolate sensitive analog blocks from the noisy digital supply. The key takeaway here is that the load characteristics affect the LDO design and must be taken into account in the design process.

An LDO can use an external, meaning off-chip, or internal load capacitor. These are the two important cases to consider separately, as they impact the entire design approach. External capacitors provide higher capacitance, but increase the cost and size of the final product. They can be used if a single LDO is used to supply a large portion of the SoC. Otherwise, if you follow the "distributed LDOs" approach, where each block is supplied by its own LDO, you will need to use smaller, internal capacitors. The choice of the capacitor essentially determines the placement of the dominant pole.

Let us first focus on the case with a large external capacitor. Here, the term "large" means higher than 1–10 nF, essentially anything higher than the values that are practical for on-chip integration, and often in the 1 µF range. A large load capacitor means that the pole coming from the LDO output will be located at low frequencies, which is why it will be selected as the dominant pole of the LDO. If the load current is sufficiently high (meaning low output resistance of the output transistor), and a smaller capacitor is used, you could use an internal dominant pole, but this will likely lead to very low bandwidth and an impractical design.

Suppose now that the dominant pole is located at the LDO output. All the internal poles must be pushed to frequencies beyond the unity-gain frequency. The main problem is coming from the gate capacitance of the large pass transistor and the pole associated with it. Ideally a low output impedance EA is needed, but achieving high voltage gain would then cost a lot of power. One strategy is to split the EA into a high-gain first stage, and a low gain, low impedance buffer that drives the pass transistor. The small input capacitance of the buffer stage should ensure that the pole of the first stage is relatively high, and its output impedance should be sufficient to push the second EA pole to a high frequency. An example circuit is given in Figure 2.8. The first stage is a simple differential pair, while the second stage is a CS stage with a diode connected transistor as a load. An alternative implementation with the same output impedance would be a PMOS source-follower. The output impedance is then equal to $1/G_{m7}$, and the bias current should be chosen such that this

Figure 2.8: Example of an LDO with a large external capacitor.

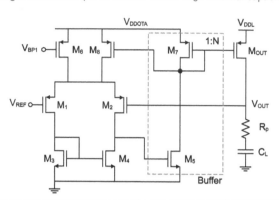

impedance is sufficiently low to push the non-dominant pole beyond the UGF. If the source-follower impedance is not sufficiently low, the buffer could be implemented as a super source follower, thereby lowering the buffer impedance using the negative feedback instead of simply increasing the buffer current [5].

Again, the difficulty lies in the fact that an LDO must be stable for a whole range of load currents. Low load current yields high output impedance of M_{out} and pushes the dominant pole to low frequencies, making the LDO more stable. Higher load current leads to an increase in $G_{ds,out}$, which increases the dominant pole frequency. The difficult part of designing an LDO with an external capacitor is making it stable for the maximum load current. Normally, you would have to make sure that the non-dominant poles are sufficiently high in the worst-case conditions, meaning that you would burn an unnecessary amount of power at low load. One way to improve the design is to use adaptive bias, this means that the bias current of the EA follows the output current, thereby pushing the internal poles to higher frequencies when the load current increases. The concept is shown in Figure 2.9. Going back to the design in Figure 2.8, note that the output stage is a current mirror, meaning that the drain current of M_7 (due to the applied negative feedback) follows the output current. Increasing I_{D7} increases G_{m7} and hence the pole frequency. Addition of M_8 does the same for the first stage. Essentially as its current increases so does its transconductance and its output conductance, pushing the pole further away. Strictly speaking, the drains of M_7 and M_8 are not at the same voltage as V_{out}, so the ratio of currents may not be exactly equal to the ratio of transistors, but normally the precision of this ratio is not crucial. Do not forget that the adaptive bias creates another loop. It is essential to assure that this loop is stable, and that

Figure 2.9: Conceptual schematic of an LDO with adaptive bias.

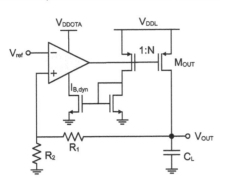

it doesn't interfere with the main loop. If needed, this can be done by adding an RC filter to slow down the adaptive bias loop.

When using external components it is important to take into account the parasitic effects. A large capacitor might have a resonant frequency in the megahertz range. This is partially due to the parasitic inductance of the component itself, but also due to the inductance of PCB traces and bond wires. These parasistics cause an increase in the output impedance past the resonant frequency, meaning that the PSRR will also increase. Some parasitic effects might be beneficial to the design. The parasitic series resistance of the capacitor (shown in Figure 2.8) creates a zero at the output. This zero can be used to null one of the non-dominant poles, and stabilize the design without the use of special techniques. The downside of this approach is that the parasitic resistance varies a lot, and rather than designing for one value, you need to ensure that the design is stable for the expected range of R_p.

Using a small internal capacitor changes the game. Typically, internal capacitors will be somewhere in the 100 pF range. If a large number of LDOs are used on a chip, it is beneficial to minimize the capacitance, and the area used for the LDOs. These LDOs are sometimes referred to as the capacitor-less LDOs, although in reality some capacitance is always present. It is simply several orders of magnitude lower than the typical external capacitance. As a consequence, the output pole will be a non-dominant pole and the dominant pole will be placed internally. As the output impedance of the pass transistor is the lowest for the lowest load current, the worst case conditions in terms of stability occur at light or zero-load current.

Since the dominant pole is internal, some form of Miller compensation (i.e. pole splitting) can always be used. If a PMOS pass transistor is used, the Miller

Figure 2.10: Miller compensation (a) and an example of an LDO with an internal capacitor and Miller compensation through a cascode (b).

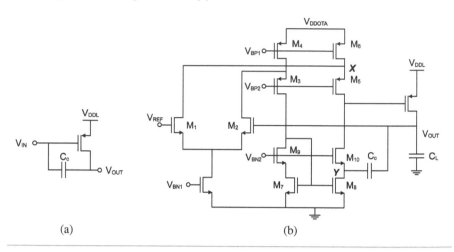

(a) (b)

capacitor can be placed directly across it. Things are a bit different with an NMOS pass transistor, in which case a Miller capacitor can be placed internally, or across the last two stages (including the negative gain stage that drives the pass transistor). Luckily, it is normally easier to compensate the LDO with an NMOS output due to its inherently lower output resistance.

For a PMOS output LDO, a single stage EA is sufficient in most cases, owing to the fact that the gain of the pass transistor is added to the loop gain. In this case a standard Miller compensation can be applied, as shown in Figure 2.10(a). By placing a capacitor across a transistor, the effective capacitance seen from the input is multiplied by the gain of that stage $C_{in} = G_m R_{out} C_c$. This allows us to use a small capacitor and to move the dominant pole to lower frequencies, and consequently improve the phase margin. The well known issue with this type of compensation is the RHP zero that appears if only a capacitor is used in the local feedback loop. A RHP zero degrades the PM if it's close enough to the UGF. To push this zero to higher frequencies, a resistor can be placed in series with the capacitor (this was illustrated in Figure 2.7). The frequency of the zero is given by

$$f_z = \frac{1}{(2\pi C_c(1/G_{m,out} - R_c))}. \tag{2.7}$$

For a sufficiently high resistor value, this will become an LHP zero, which improves the PM. The difficulty with the LDOs is that the pass transistor transconductance changes with the load current, and with it the position of the zero. A common way to remove the zero is to cut the feed-forward path

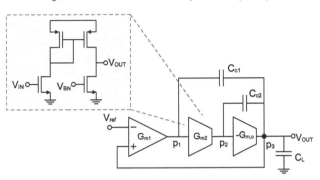

Figure 2.11: Nested Miller compensation principle.

by using a buffer. In practice this can be achieved by closing the feedback through a cascode transistor, which effectively acts as a common gate stage. This is illustrated in Figure 2.10(b), where a folded cascode stage is used as EA. Note that the desired effect can be achieved by connecting the capacitor to either node X or Y. In the shown implementation the transistor M_{10} acts as the buffer, but note also that there will be a resistance to ground created by M_8. This path creates an LHP zero that can be used to improve stability and increase the amplifier bandwidth. The resistance is controlled through the bias current, so unlike the compensation with a series resistance this zero is fixed and doesn't depend on the load current. It is also possible to bias the transistor M_8 in the linear region to further decrease resistance to ground if needed.

Adaptive bias can also be applied to LDOs with an internal load capacitance. Without the adaptive bias, you are obliged to stabilize the LDO in the worst case conditions, that is at the lowest possible load. This leads to low bandwidth of the EA, which results in a bad PSRR in the medium frequency range. Before resorting to the adaptive bias, it might be useful to try a simpler trick. Namely, it is possible to add a small dummy load current, which guarantees a minimum output resistance and relaxes the stability constraints, but increases consumption. If this is not a significant burden on efficiency it might be sufficient to provide good stability. Otherwise adaptive bias can be used, allowing us to extend the EA bandwidth for higher load currents and improve the PSRR.

Sometimes, for example when high DC gain is needed, multiple stages might be necessary for the EA. More complex compensation techniques, such as the nested Miller compensation (NMC), will be required in such cases. The fundamental way of thinking is still the same: generate one dominant pole, and move all others above the UGF (i.e. GBW of the first stage). In addition, care is required to separate the non-dominant poles sufficiently to avoid amplitude peaking near the UGF. The concept of NMC is shown in Figure 2.11. Note that the second stage needs to have positive gain, and hence a positive transconductance. It is not possible to implement such a stage with a single transistor. It can be implemented either using a current mirror (shown in the figure) or a differential pair. The approximate expressions for the GBW of the first stage and frequencies of the non-dominant poles are given by:

$$GBW = \frac{G_{m1}}{2\pi C_c 1}, \; f_{p2} = \frac{G_{m2}}{2\pi C_{c2}}, \; f_{p3} = \frac{G_{m3}}{2\pi C_L}. \tag{2.8}$$

The compensation capacitors are then chosen to have sufficiently high f_{p1} and $f_p 2$, and sufficient separation between them. Other techniques exist that can slightly extend the compensated bandwidth [6, 4]. They aim primarily to split the non-dominant poles or to compensate some of them using LHP zeros. Still, without the use of dynamic biasing, the bandwidth of the amplifier will be heavily limited by the worst-case output pole location.

We mentioned previously that a choice of pass transistor impacts the PSRR and that the NMOS pass transistor has a clear advantage. Another element that plays an important role is the load capacitor. Using a big external capacitor or a small internal one makes a big difference. The typical PSRR curves for the two cases are shown in Figure 2.12. The low impedance of the external capacitor will often be sufficient to provide good supply rejection. At very low frequencies the PSRR is dominated by the gain of the EA. This is the case until the point where the low impedance of the capacitor takes over. When using an external capacitor it is important to pay attention to parasitic elements. Assume you're using a 1 μF capacitor. A combination of capacitor series inductance, PCB trace and a bond wire could easily lead to a parasitic inductance in the order of 1 nH or more, resulting in a resonance frequency of approximately 5 MHz. Beyond this frequency the load capacitor actually starts to behave as an inductor, which means its impedance will increase with frequency. Whether it is important or not, depends on the frequencies of interest and the desired supply noise suppression, but should definitely be considered when designing an LDO.

If an on-chip load capacitor is used, the dominant pole will come from the EA. This means that the EA bandwidth is lower than the frequency at which the load capacitor impedance is low enough to attenuate supply noise. Unfortunately it is not possible to simply increase the load capacitor as this would bring the non-dominant pole close to the UGF. We can distinguish three separate regions in the PSRR curve. At low frequencies it is dominated by the EA. At very high frequencies the PSRR is determined by the ratio of the drain-source capacitance of the pass transistor C_{DS} and the load capacitor C_L, i.e. the PSRR is entirely determined by this capacitive divider. In between

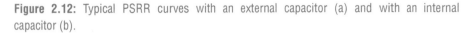

Figure 2.12: Typical PSRR curves with an external capacitor (a) and with an internal capacitor (b).

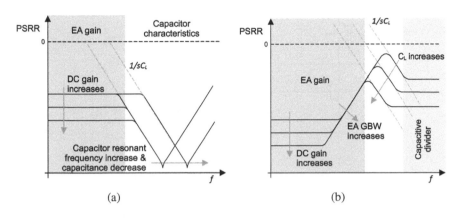

(a)   (b)

these two regions the shape depends on several factors. Some intuition can be gained by looking at Equations 2.5 and 2.6 (but note that C_{DS} was neglected in these equations). To simplify, assume that the unity gain feedback is used, that $R_L \to \infty$, and that we are looking at frequencies where EA gain is small, which is close or above the UGF. Equation 2.5 becomes

$$\frac{V_{out,CS}}{V_{DDL}} \approx \frac{G_m + G_{ds}}{G_{ds}(1 + sC_L/G_{ds})}. \tag{2.9}$$

We can conclude that there is a sort of plateau roughly equal to G_m/G_{ds}, i.e. intrinsic gain of the common source stage; this is the maximum that the PSRR can reach, and as we can see it is above 0. This is a very intuitive result given that the source of the pass transistor is directly connected to the supply. Whether this value is reached or not depends on the EA bandwidth and C_L. Note that the cut-off frequency is in fact the non-dominant pole frequency. For a sufficiently high C_L the plateau may never be reached, allowing to keep the PSRR below 0, but note that stability may be affected. Another way to reduce the PSRR peaking is the feedforward cancellation using a series connection of a resistor and a capacitor at the PMOS gate, as shown in Figure 2.13(a). The idea is to place a positive zero (from the perspective of supply V_{DD}) directly on top of the peak [7], and lower it in that way.

Using the same assumptions, from Equation 2.6 for the SF LDO we get

$$\frac{V_{out,SF}}{V_{DDL}} \approx \frac{G_{ds}}{G_m(1 + sC_L/G_m)}. \tag{2.10}$$

Figure 2.13: Different combinations of OTA and pass device for improved PSRR.

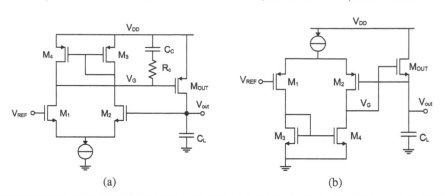

(a)   (b)

We can see that in this case the PSRR should always remain below 0, as the maximum is now G_{ds}/G_m. The same conclusion holds for C_L as before. It's interesting that in the SF case you can improve the PSRR by increasing the pass transistor output resistance, while the opposite is true for the CS LDO. Equations 2.5 and 2.6 are also valid for the case with the off-chip capacitor. The key difference is that C_L determines the dominant pole here, so you are free to choose a very large capacitance and avoid the peaking region entirely.

So far we've considered only the pass transistor while analyzing the PSRR. While this is the dominant path, it's also worth considering the EA, as the choice of OTA topology has an impact on the overall PSRR. To illustrate this let us focus on the two basic topologies in Figure 2.13. For the OTA with a PMOS current mirror in Figure 2.13(a) it can be shown that the entire supply ripple appears at its output V_G [8], which is the gate of the pass transistor. It is a good idea to combine this OTA with a PMOS pass transistor, as ideally the gate and source voltage will cancel out and no supply ripple will be coupled to the output. For the OTA with an NMOS current mirror from Figure 2.13(b) no supply ripple will be coupled to the output in the ideal case. In reality there will always be at least some coupling due to mismatch, but this will likely not be the dominant path. Pairing this type of OTA with an NMOS pass transistor should lead to good PSRR, as the OTA itself shields the gate of the pass device. Note that this configuration may lead to difficulties with the output voltage range.

For a couple of final remarks, remember that the reference PSRR must also be considered, as the LDO behaves as a buffer from its perspective. This means that the low-frequency noise from the reference translates directly to the LDO output. Improving the LDO PSRR will be meaningless if a bad reference is used.

The PSRR issue can also be approached from different angles. For example you could cascade the LDOs, although this is rarely done in practice as it will increase the dropout voltage. From the load perspective, the PSRR can be built into the circuit supplied by the LDO, for example by using a differential circuit. This could relax constraints on the LDO and simplify its design. It is sometimes worth considering the full system as it might lead to a globally more optimal solution than just optimizing a single block.

2.5 Noise

The unwanted signal at the LDO output comes from the two sources. The first is the supply noise coupling to the LDO output. We call it noise although it might also be an unwanted deterministic signal, such as the supply ripple of a DC–DC converter. Suppression of this noise is quantified by the PSRR and it was discussed in the previous section. The second source is the noise of the LDO itself. Different sources of noise in an LDO are shown in Figure 2.14.

From the figure, we can calculate the contribution of each source to the output noise. The output noise power spectral density (PSD) is given by

$$S_{Vnout} = \left(1 + \frac{R_1}{R_2}\right)^2 \left(S_{n,ref} + S_{nEA} + \frac{S_{nMout}}{A(s)}\right) + S_{nR2} \left(\frac{R_1}{R_2}\right)^2 + S_{nR1}, \quad (2.11)$$

where the PSDs of different noise contributors are denoted by S_n. Total (flicker and thermal) noise should be accounted for in all the noise contributors. The output nose voltage (actually its square) is obtained by integrating S_{Vnout}. For

Figure 2.14: Noise sources in an LDO.

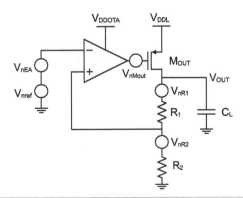

simplicity, any compensation circuits, as well as the load capacitor are left out of the calculation. However, it is important to remember that they will have an impact on the shape of the noise PSD, especially the load capacitor that will attenuate all the high-frequency components.

The first important thing to notice is that the noise of the pass transistor is attenuated by the EA gain. This is an expected result, as the action of the negative feedback counters any unwanted disturbances. Just like with the PSRR, this will be the case inside the bandwidth of the EA, that is as long as $A(s)$ is larger than 1. The PSD S_{nMout} is generally not expected to be the dominant factor at low frequencies. Outside the EA bandwidth, however, M_{out} is directly injecting noise into the output node. Furthermore, this noise is proportional to the output current, so it is likely to be the dominant factor in the mid–high frequency range, especially at high load. The EA noise and the reference noise are directly present at the output, again an expected result given that from the perspective of the reference terminal, the LDO as a whole acts as a buffer with a gain defined by the resistive divider and equal to $(1 + R_1/R_2)$. In fact, all three of the mentioned noise components will be amplified by the same gain factor. Reference noise, if it poses a problem, can be filtered, but this comes at the price of increased area and start-up time. The EA noise is likely to be dominant at low frequencies. It is the sum of contributions of all the OTA transistors, although it is normally dominated by the noise of the first stage which is amplified by all the subsequent stages. For a standard OTA, such as the one in Figure 2.6(a), the noise contributions from the differential pair and the current mirror are equal. If flicker noise is an issue, a possible way to reduce it is to increase size of the differential pair and the current mirror (as a bonus this also improves matching). Another way to decrease output noise is to increase the transconductance of the first stage. This actually increases the output noise of the EA, as the MOS transistor noise current is proportional to G_m. However, when referred to the input, the EA noise contribution is divided by G_m^2, resulting in a net decrease in the LDO output noise. Resistor noise is also directly visible at the output. This contribution can be decreased by using smaller resistors, at the cost of increasing output current. Placing a capacitor across R_1 reduces the noise contribution of both resistors, but may impact stability and dynamic behavior of the LDO. Still, if possible, the simplest way to avoid the added noise due to resistors is to remove the resistors and use the unity feedback configuration. This is also beneficial for $S_{n,ref}$, S_{nEA} and S_{nMout}, as the gain coming from the resistive divider is gone. Although an LDO with a common source pass transistor was used for the calculation, the results are also valid for the source follower output. In this regard there is practically no difference between the two types of LDO.

Figure 2.15: Undershoot enhancement using AC coupling.

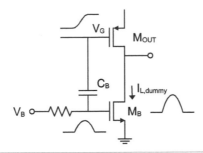

Figure 2.15: Undershoot enhancement using AC coupling.

2.6 Useful LDO Techniques

This section provides an overview of some less standard techniques, which may prove to be useful in some cases. These are the circuits and techniques you might resort to if a more conventional approach doesn't work, or if you want to optimize the LDO for a very specific scenario.

When it comes to transient response, there is one tricky case that might be of concern. This is the case when the load current suddenly drops to 0, which essentially corresponds to switching off the block supplied by the LDO. This scenario triggers the worst-case overshoot, which could exceed the technology limits and damage the circuits. The usual way to combat this is to increase the bandwidth of the EA, in order to speed up the reaction time of the feedback, or to increase the load capacitance. Either of these approaches is limited by stability requirements. Another issue is that once the load is completely off, the load capacitor will be very slowly discharged through feedback resistors, or through capacitor leakage if a unity-gain feedback is used. A simple way to solve the problem is shown in Figure 2.15 [9]. The solution is to add a parallel loop with a bandpass characteristic that will react quickly to counter the perturbation in the output voltage. The transistor M_B also provides a path that can quickly discharge the output capacitor, and can be biased to draw a small dummy current and improve the PM of the LDO in steady state. Aside from the added load capacitance in the V_G node, there is no additional impact on the stability when applying this technique. The technique can be applied differently, by using AC coupling in different points (for example AC coupling from the output node [10, 11]), and alternatively to enhance the driving capability in the node V_G instead of directly injecting current at the output.

An interesting topology is the flipped voltage follower (FVF), shown in Figure 2.16(a). It is particularly interesting when combined with an on-chip capacitor, as the dominant pole is placed at its output, allowing significantly higher bandwidth compared to standard integrated LDOs with an internal dominant pole. This further enables a faster transient response and a good mid to high frequency PSRR without peaking. For the simple implementation in Figure 2.16(a) the loop gain is given by

$$A_{LG} = \frac{\frac{G_{m,out}G_{m1}}{G_{ds,out}G_{ds1}}}{1 + s\frac{C_G G_{m1}}{G_{ds,out}G_{ds1}} + s\frac{C_L}{G_{ds,out}} + s^2\frac{C_G C_L}{G_{ds,out}G_{ds1}}}, \tag{2.12}$$

where C_G is the gate capacitance of the pass transistor. One downside of the simple topology is the relatively high impedance seen from the gate of the pass transistor, which then limits the non-dominant pole frequency. This is why the implementation of Figure 2.16(b) is more commonly used in practice. The super-source follower that consists of transistors M_2 and M_3 acts as a buffer and reduces the impedance seen by C_G, pushing the non-dominant pole to higher frequencies. The only downside compared to the simple version is the reduced voltage headroom for the gate of the pass transistor. The FVF is itself a feedback system and, assuming the correct reference voltage is applied at the input, it will provide a regulated output. More commonly, a second feedback is provided using an OTA, in which case you can simply see the FVF as an output stage with a unit gain. There are two options here, the first is to keep the internal poles high, by using limited gain, and maintain the dominant pole at the output. The second is to use a very low frequency OTA, with a relatively large capacitor at the input of the FVF. In the second case the slow feedback loop through the OTA essentially provides a reference voltage for the FVF, while the fast FVF loop assures fast transient response and a good high frequency PSRR. In the implementation from [5] a combination of the two loops through the OTA is used (for a total of three loops), providing at the same time good regulation and fast response time.

We've mentioned previously that the stability, EA bandwidth, PSRR and load capacitance are related and that you cannot change one without affecting the others. A way to decouple them is to use replica biasing [12]. The principle is shown in Figure 2.17. Since the replica branch is used to close the feedback, the loop stability can be guaranteed independent of the load. It is now possible to use a large load capacitor without worrying about the non-dominant pole. The output voltage is guaranteed by matching the ratio of M_B and M_{out} to the ratio of I_B and I_L. The concept can be extended to multiple outputs while using a single EA. Obviously, this is only possible if the load current of each supplied block is well defined, and limited to a relatively narrow range. Since the gate voltage of M_{out} doesn't depend on the load, the output voltage will be

Figure 2.16: Flipped voltage follower in basic configuration (a) and using a super source follower (b).

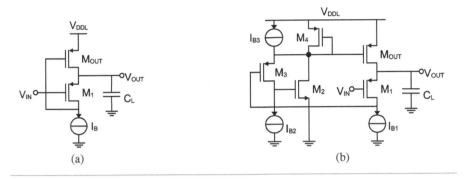

(a) (b)

Figure 2.17: LDO with a replica bias.

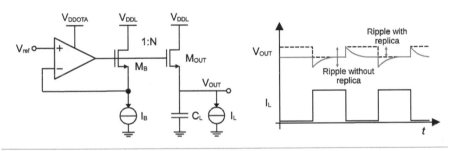

determined by the I–V characteristic of the source follower; for small changes in current the voltage shift will be $\Delta V_{out} = \Delta I_L / G_{m,out}$. That is, the output impedance is no longer affected by the negative feedback and depends purely on the source follower. Interestingly, the output ripple is lower than without the replica bias, as shown in Figure 2.17. If a current pulse is applied to the standard LDO, the output voltage will first drop by an amount equal to $\Delta I_L / G_{m,out}$, then, the feedback will bring it back to its desired value, the opposite happens when the current drops. As a result, the equivalent ripple is approximately two times higher without the replica bias, but the steady-state voltage is practically constant.

There are some obvious downsides of the replica bias approach. First of all, the proposed topology cannot be realized using a PMOS pass device due to its high output resistance. In order for this topology to be efficient you either need to have the higher voltage for the OTA present somewhere in the system or you need to use a charge pump to generate it. The PSRR will be determined by the

Figure 2.18: Digitally controlled LDO and output waveform.

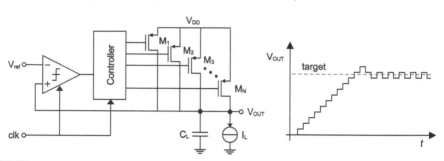

supply rejection of the source follower as the output transistor is no longer part of the feedback. The good news is that you can avoid the peaking in the PSRR as stability is no longer an issue, the bad news is that you no longer benefit from the high suppression coming from the EA gain at low frequencies.

Modern CMOS technology nodes are optimized for digital circuits, so it is only natural that the tendency is to move more and more functions into the digital domain. A digitized version of an LDO is shown in Figure 2.18. One of the advantages of digital circuits is the size, for example digital filters implemented as delays, additions and multiplications are often significantly smaller than their analog counterparts implemented with resistors and capacitors. Another advantage is the ability to operate from relatively low voltages. The design in [13] can work from 0.45 V. Indeed, implementing an OTA that operates from such a low supply would be very challenging. Instead, the digital LDO uses a clocked comparator that provides the information on whether the output voltage is higher or lower than the reference. The pass transistor is split into a large number of small sections turned on and off by the digital controller. The output voltage is regulated by turning on the appropriate number of output transistors. Depending on the desired behavior, the digital controller can be more or less complex, and the input could be quantized using multiple levels (requiring multiple comparators). The implementation from Figure 2.18 uses a very simple controller that turns on an additional PMOS if the output is lower than reference, and turns a PMOS off if it is above the reference. An obvious downside of this approach is that for a constant current the output will oscillate around a fixed value. By using a sufficient number of output transistors and a sufficiently large capacitor the output ripple can be reduced to the desired value. The CMOS digital circuits ideally consume no static power (in reality there is always some leakage current), so the consumption of the digital LDO is proportional to the clock frequency. At the same time the maximum speed at which the LDO can react is determined by the clock and the output step size. As for the PSRR, the loop is only able to suppress perturbations below the clock frequency, above it the LDO will behave as a resistive or a capacitive divider. Just like in the analog circuits, we see that we also have a trade-off between bandwidth and the PSRR on one side and consumption on the other. The stability remains a concern, the digital controller has poles (that also depend on the clock) and their placement must be considered together with the pole coming from the output capacitor.

There are many ways in which the basic implementation shown here can be improved. It is possible to eliminate the output ripple by adding another comparator, so that the output remains constant within a certain range. If the output voltage moves above or below this range the LDO will react. An adjustable clock can be used to provide higher speed at higher load currents (the concept is similar to adaptive bias). Multiple loops can be used with different output steps, e.g. a coarse loop that provides high output current and a fine loop that provides higher regulation accuracy [14]. Hybrid LDOs have been proposed [7] as a parallel combination of an analog and digital LDO, where the role of the analog part is to improve the PSRR and enable reduction of the output capacitor. At the same time a large portion of the output current should be handled by the digital LDO, allowing the minimization of the quiescent current. Finally, an interesting combination of analog and digital is a class D LDO that uses pulse-width modulation (PWM) to drive the output transistor. As the output is not directly switched the ripple is lower than in a standard digital LDO implementation, and it still keeps the low quiescent current and small silicon area as found in the purely digital implementations. One downside is that a full OTA is required for the EA, which limits the minimum supply voltage.

3

Switching Converters

Switching converters are used for efficient voltage conversion from one DC level to another. The resistive drop that is inherently present in an LDO implies losses, especially as the difference between the input and output voltage increases. Unlike the LDOs, the switching converters do not rely on resistive voltage division. Instead they use switches and ideally lossless reactive components, i.e. inductors and capacitors, to manipulate the average values of the input and output voltage. From a very high-level perspective, they convert the DC voltage into AC and back, performing efficient voltage conversion in the process. Also unlike the LDO, the switching converters have the capability to convert the low input voltage into a high output voltage, making them useful for battery charging and energy harvesting applications.

In this chapter we will look into the analysis and design of the switching converters. As it is often the case, we will start from the steady state and explain how to optimize efficiency and size the switches of the converter, a key step in the design procedure. Next, the dynamic behavior of converters is analyzed in order to explain how to design the control circuits, from a high-level block diagram down to transistor level implementation. Since we are primarily targeting battery powered SoC, an overview of low-power techniques is given, which prove to be useful for extension of the battery life. Capacitive converters are also covered in this chapter. Although they are not as efficient as the standard inductive DC–DC converters, they are smaller and simpler and often used as an auxiliary supply on the chip.

3.1 Switching Converters for Low-power Integrated Systems

There is a wide range of developed converters with different characteristics, driving capabilities, number of components, galvanic isolation, etc., covering many different applications [15]. In the context of low-power systems-on-chip, we will talk about two types of switching converters: (inductive) DC–DC converters and switched-capacitor (SC) converters. Again, there are many different topologies of DC–DC converters, but when a high level of integration and low power consumption are the priorities, there are some specific constraints.

Since the aim is to integrate the converter together with the system, the silicon area is important, meaning that the simpler topologies are preferred. Typically, the largest components in the converter are the switches, as they carry most of the current and must present a low on-resistance. From that perspective it is good to use a minimum number of switches. Another important factor is the number of off-chip components. Obviously, we want to reduce this number as much as possible, as this allows for a smaller and cheaper system. For battery-powered devices galvanic isolation is not needed, so you will rarely see an integrated DC–DC converter with a transformer. From these observations it is clear why the two basic converters, buck or step-up and boost or step-down converters, are almost exclusively used in low-power SoCs. The third basic converter, buck-boost, is rarely used in this context as it inverts the output voltage. A version that doesn't invert the output exists, but it requires more switches, which increases the silicon area, and degrades efficiency. In addition, you'll rarely encounter a case in practice where the output voltage may need to cover the range above and below the input voltage.

The three basic converters and the non-inverting boost converter are shown in Figure 3.1. They all need at least two large reactive components, commonly placed outside of the chip: an inductor and a capacitor. A second large capacitor is usually placed for input voltage decoupling. Usually the inductor is in the 1–50 µH range, while the capacitors take values in the 1–50 µF range, making both of them highly impractical for integration. Converters with smaller, integrated passive components have been implemented, but at the cost of significant loss of efficiency and performance in general, mainly due to the low Q-factor and small inductance of the integrated inductors.

In all of the converters from Figure 3.1 diodes are used along with actively controlled switches. When input and output voltages are relatively high compared to the diode threshold voltage, this is not an issue as the impact on efficiency will be negligible. However, for a battery powered SoC, where the

Figure 3.1: Schematics of basic converters: buck (a), boost (b), buck-boost (c) and non-inverting buck-boost (d).

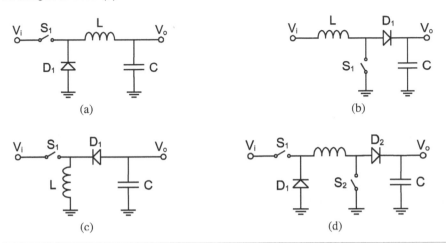

(a)

(b)

(c)

(d)

voltages typically take values in a 1–5 V range, a voltage drop of 0.7 V across a diode would be the dominant source of losses. This is why in the integrated version, the diode is replaced by a synchronous MOS switch. This allows the efficiency to be improved, but again increases complexity. A diode is a passive switch, meaning that its state is controlled by the applied current and voltage. A diode conducts as long as its current is positive and acts as an open circuit when the current reaches 0. A MOS switch needs to be switched off by external circuitry, which also requires monitoring of its drain current.

The main challenge when implementing an integrated converter for a low power system is maximizing efficiency for low load currents. By low we mean somewhere in the range from a few microamps up to several hundreds of milliamps. SoC will almost always have different modes of operation. Most of the time they will be in sleep mode, where only the essential parts of the chip are kept on such as a small part of memory to conserve the CPU state, a real-time clock or peripherals used to wake up the system. This is necessary to achieve a decent battery life, and the power management itself must not consume more than a small fraction of the SoC consumption in this low power mode. In these cases the controller of the DC–DC converter becomes the bottleneck and the main design challenge is to optimize its consumption. When the SoC is drawing current in the order of milliamps, the controller consumption is often negligible, and instead the switches and passives must be optimized, a case typically described in the literature [15]. Just like the SoCs, the converters will

Figure 3.2: Capacitive voltage doubler.

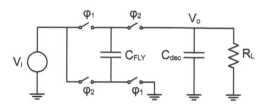

also use different modes of operation, optimized for different load currents, aiming to provide high efficiency in all scenarios.

In essence the DC–DC converters perform a similar role to the linear regulators, they just do it more efficiently. Most of the parameters relevant for the linear regulators are also relevant here. The line and load regulation, transient behavior, stability, PSRR, output impedance, etc. One difference is that we no longer worry about noise, instead we are concerned with the output ripple that appears due to the switching activity. The second important difference is that the efficiency is not determined by the difference in input and output voltage (at least in the first order approximation). The main purpose of DC–DC converters is to provide efficient conversion for relatively large voltage ratios. If the input voltage is close to the output an LDO might prove to be a more efficient solution due to its simplicity.

As the name would suggest, SC converters only use switches and capacitors. Figure 3.2 shows a simple voltage doubler (or a charge pump), where the C_{FLY} is first charged to V_i, and then in the second phase the capacitor voltage is added to the input voltage. Once the C_{dec} is fully charged, and if there is no load current, the output will be at exactly twice the input voltage, but decreases as the load current is applied.

This decrease depends on the output impedance of the charge pump, which is a function of the capacitances and the switching frequency. In general, the higher the capacitance and the switching frequency, the lower the output impedance. If a charge pump is used to supply circuits that draw relatively low current, small capacitors, practical for on-chip implementation could be used. Unlike the DC–DC converters, SC converters can be fully integrated, however their efficiency is typically lower and degrades as the load current increases. Aside from the given example, different configurations of switches and capacitors can be used to provide different multiplication ratios of the input and output voltage. These ratios can be above and below 1 as well as negative, but are always fixed for a given configuration. If different values of

Figure 3.3: Buck converter basic schematic.

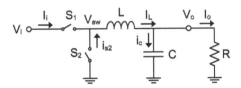

the output voltage are needed, configurable SC converters can be implemented that can support several different ratios and combined with switching frequency control to provide a continuous voltage range. In practice the SC converters are used to supply a small part of the system that requires a specific voltage. A typical example would be an LDO with a source follower, whose error amplifier requires a higher voltage than the core voltage, or they can be used to provide the negative voltage for the body bias in order to minimize leakage of NMOS transistors in the off state.

3.2 DC–DC Converters in Steady State

The analysis of DC–DC converters usually starts with the steady state, where we assume that the input and output voltage and load current are constant. Looking at the steady state provides insight into the operation of DC–DC converters and allows identification of dominant loss mechanisms. This where the design starts as well, as the first step is to size the switching transistors and choose the appropriate passive components. The choice is driven by different requirements, such as the maximum output ripple, efficiency, switching frequency or the footprint of the external components. In order to understand how these quantities affect one another we will now look into the steady-state operation of the buck and the boost converter.

A buck converter is shown in Figure 3.3. As we mentioned previously, in a buck converter integrated in a CMOS technology, a diode is replaced by a synchronous switch. When the switch S_1 is closed the inductor sees V_i and when the switch S_2 is closed it sees 0. If the switches close and open alternately at the switching frequency f_s, which is much higher than the cut-off frequency of the LC filter, at the output we will get the average value of the voltage V_{sw}. Assuming that the switch S_1 is on for a period of time DT_s, where D is the duty cycle and $T_s = 1/f_s$ is the switching period, the output voltage will be equal to DV_i. This is a simple, intuitive way to understand the buck converter; however, a bit more formalism is needed to analyze the steady state.

When analyzing DC–DC converters a common assumption is that the capacitor is sufficiently large that its voltage can be considered constant. By doing this we are essentially neglecting the output voltage ripple and assuming that the inductor voltage is constant for a fixed state of the switches. The second assumption is that all elements are ideal, i.e. for now we ignore the losses in the circuit. When the switch S_1 is closed, the inductor voltage is $V_i - V_o$ and so the current through the inductor increases linearly. When the S_1 opens and S_2 closes, the inductor voltages changes to $-V_o$, and the current now starts to decrease until the end of the switching period when the whole process repeats. The voltage and current waveforms are illustrated in Figure 3.4(a). Necessary tools for analyzing the steady-state are the volt-second balance, which tells us that the average voltage across the inductor is 0, and the ampere-second balance, which tells us that the average current into the capacitor is 0:

$$\int_t^{t+T_s} v_L(t)\, dt = 0, \quad \int_t^{t+T_s} i_C(t)\, dt = 0. \tag{3.1}$$

From the volt-second balance on the inductor we easily obtain the output to input ratio of the buck converter as $M = V_o/V_i = D$. We can see that the converter acts as an ideal DC transformer. Average current through the inductor I_L is equal to the output current I_o. The two slopes of the inductor current depend on the input and output voltage and the inductance, knowing this and the duty cycle we can calculate the current ripple (one half of the peak-to-peak value) as

$$\Delta I_L = \frac{(1-D)D}{2Lf_s} V_i. \tag{3.2}$$

Based on this equation, the inductance and the switching frequency can be chosen such as to reduce the current ripple to the needed value.

It is also possible to estimate the capacitor ripple based on the previous reasoning. As long as the ripple is relatively small compared to the output voltage, all the provided results will be valid. The capacitor ripple has two components, the first one comes from the charge fluctuation on the capacitor and can be calculated by integrating the capacitor current. For example, the current can be integrated in the half-period when it is positive to obtain the peak to peak value of the capacitor voltage ripple

$$\Delta V_C = \frac{\Delta I_L}{8Cf_s}. \tag{3.3}$$

The second component comes from the equivalent series resistance (ESR) of the capacitor and is equal to $R_{esr}i_C(t)$. The two components can be combined to calculate the capacitor voltage over time as

$$v_C(t) = V_C(t_0) + \frac{1}{C}\int_{t_0}^{t_0+t} i_C(t)\, dt + R_{esr}i_C(t), \tag{3.4}$$

Figure 3.4: Buck converter waveforms in steady state CCM (a) and DCM (b).

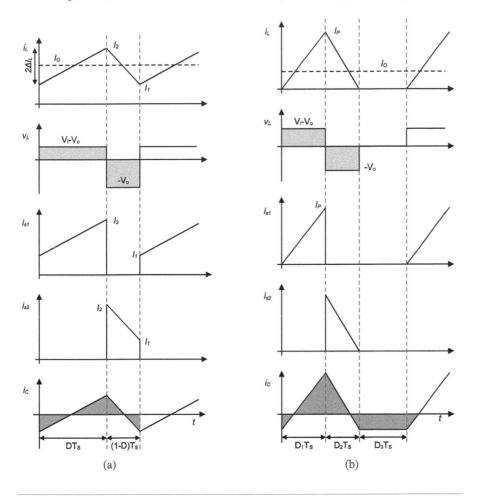

(a) (b)

although it is quite common that the capacitor voltage ripple is entirely dominated by the component coming from the ESR, making the output voltage ripple directly proportional to the current ripple.

So far it was assumed that the inductor current was always positive, this is known as the continuous conduction mode (CCM). What happens if the current reaches 0? In the implementation with the two switches, nothing; the current will simply change its direction and CCM equations will still be valid. In the classical implementation with a diode, as soon as the inductor current reaches 0, the diode ceases to conduct and disconnects the V_{sw} node. A third part of

Figure 3.5: Buck converter switch implementation (a) and V_{sw} waveform in DCM (b).

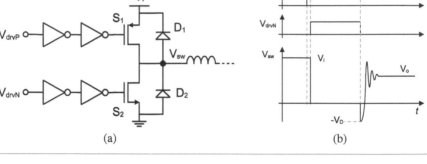

(a)   (b)

the switching period appears with both switches open and no current flowing through the inductor. This is known as the discontinuous conduction mode (DCM). In the integrated implementation the switch S_2 is opened as soon as the zero current is detected to avoid discharging the capacitor through the inductor. In other words, the DCM is forced in order to avoid a loss in efficiency. An implementation of the buck converter with MOS switches and drivers is shown in Figure 3.5(a). In standard bulk technologies, parasitic diodes will always be present in parallel with the two switches; these diodes are the PN junctions between the bulk and the drain diffusion for the NMOS, and the n-well and the drain diffusion for the PMOS. In DCM, when the switch S_2 is open, one of these two diodes will discharge the inductor current. In the case shown in Figure 3.5(b) the switch is open early and since the inductor current is positive, the diode D_2 will turn on forcing V_{sw} to $-V_D$, which is the diode voltage drop, until the inductor current reaches 0. At this point V_{sw} jumps to V_o. Some ringing will appear due to the resonance caused by the inductor and the capacitance in the node V_{sw} (parasitic drain capacitance of the switches and the IO pad capacitance), but this ringing should have no impact on the average currents and voltages, meaning the introduced approximations are still valid. If the switch S_2 is opened late, the inductor will discharge through the diode D_1, forcing the V_{sw} briefly to $V_i + V_D$ before settling at V_o. The conducting diodes will cause some losses in the circuit, but these are typically small compared to the conduction losses in the switches and the inductor.

The idealized buck converter waveforms in DCM are shown in Figure 3.4(b). The simple CCM relation between input and output voltage is no longer valid. Note that there are now three distinct parts of the period. Duty cycle D_1 of the switch S_1 is controlled, but D_2 is unknown and depends on the system parameters and load resistor (or current). Volt-second balance can be used to

Figure 3.6: Boost converter basic schematic.

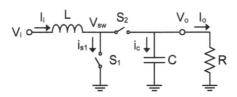

determine the relation between input and output voltages and D_1 and D_2. Using the equations for the inductor current slope, and relation between the load current and the peak current $\frac{1}{2}(D_1 + D_2)I_p = I_o$, we can calculate the output to input ratio as

$$M = \frac{V_o}{V_i} = \frac{2}{1 + \sqrt{1 + 4K/D_1^2}}, \tag{3.5}$$

where $K = 2L/RT_s$. The expression changes if the load resistor is replaced by a current source, in fact in that case it will not be possible to get the explicit V_o/V_i ratio as a function of other parameters. The key difference compared to the CCM is that the output voltage now depends on the duty cycle D_1 and the output load (in this case a resistor R). The boundary between the CCM and DCM can be found from the condition $2\Delta I_L = I_o$, which results in

$$I_o = D(1 - D)\frac{V_i}{2f_s L} = I_{crit}. \tag{3.6}$$

When the load current is above I_{crit} the converter operates in CCM, and below it in DCM. Note that, aside from the factor $f_s L$, the critical current also depends on D, that is, on the ratio of the output and input voltage.

Similar analysis can be conducted for the boost converter using the same assumptions. The boost schematic is shown in Figure 3.6. In the first phase the switch S_1 is closed and the inductor current grows linearly. Essentially, during this phase energy is stored in the magnetic field of the inductor. In the second phase, when the switch S_2 closes, the stored energy is transferred to the output. Over this period the inductor current decreases. All the relevant CCM waveforms are shown in Figure 3.7(a).

In a similar manner, using the volt-second balance, we can obtain the ratio of the output and input voltage $M = V_o/V_i = 1/(1 - D)$ in CCM. The inductor current ripple for the boost converter is given by

$$\Delta I_L = \frac{D}{2Lf_s}V_i. \tag{3.7}$$

In this case the average current through the inductor is equal to the input current, while the output current is $I_o = (1 - D)I_i$. The current ripple of the

Figure 3.7: Boost converter waveforms in steady state CCM (a) and DCM (b).

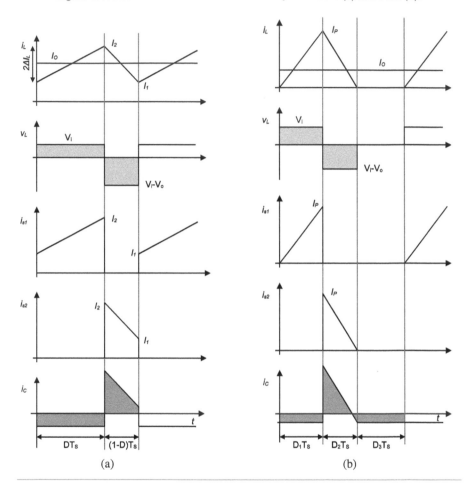

(a)   (b)

capacitor current is now higher and equal to the peak of the inductor current, which translates directly to the output voltage ripple, assuming the ESR is the dominant factor. For a low ESR value the ripple can be calculated by noting that in the part of the period when the switch S_2 is off, a constant current I_o is drawn from the capacitor, so the ripple amplitude must be be $\Delta V = \frac{DT_s I_o}{2C}$.

A small digression should be made here, in relation to the voltage ripple. Aside from the parasitics of the passive elements, it is important to be aware of the parasitics at the chip interface. By this we primarily mean the inductance and resistance of the bond wires, used to connect the chip pads to the PCB or the package substrate. As an example a boost converter schematic is shown in

Figure 3.8: Boost converter with off-chip parasitics.

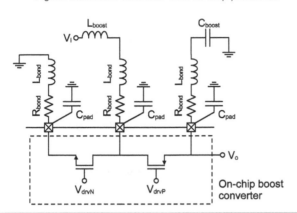

Figure 3.8 together with resistance and inductance of the bond wires. Elements in series with the boost inductor won't have much effect, other than a small impact on efficiency. But, as the boost capacitor is connected to the output through a small inductance, the capacitor current discontinuity results in a short spike in the output voltage. This is due to the resonance caused by L_{bond} and C_{pad}. Note that this voltage will not be visible directly on the capacitor, but only inside the chip. The spikes can be reduced by using multiple pads for V_o and and several bond wires in parallel, but never completely eliminated. Additional on-chip decoupling capacitors or LDOs are needed to filter out these spikes and prevent them from polluting the supply of sensitive circuits.

Just like in the buck converter case, if the load current is low enough, the inductor current will drop to 0 during the switching interval, and the converter will operate in the DCM. The input-output relation must be derived separately for the waveforms presented in Figure 3.7(a). The volt-second balance can be used to deduce relation between D_1 and D_2 as $D_1 V_i = (V_o - V_i)D_2$. Combining this expression with the current slope equations and the relation between the peak current I_p and the output current $I_o = \frac{1}{2}D_2 I_p$ we obtain the output to input ratio as

$$M = \frac{V_o}{V_i} = \frac{1 + \sqrt{1 + 4D_1^2/K}}{2},\qquad(3.8)$$

where again $K = 2L/RT_s$. The critical current for the DCM–CCM boundary of the boost converter is given by Equation (3.6), interestingly it is equal to the critical current of the buck converter. The important thing to note here is again the output voltage dependency on the load.

3.2.1 DC–DC Converter Losses

So far we have assumed that converters were ideal, meaning that there are no losses. In reality, a part of the input power will be lost in the converter. In order to optimize the switches and the passive components it is important to understand the different loss contributions. Usually the most significant losses are coming from the switch resistance and the series resistance of the inductor. Losses on the ESR of the capacitor are generally negligible.

One way to estimate losses is to use the zero-ripple approximation. This means neglecting both the capacitor voltage ripple and the inductor current ripple (in the derivations above the linear-ripple approximation was used which takes into account the ripple of the inductor current), and simplifies the calculation. The volt-second balance, ampere-second balance, and power balance equations can then be used to derive the modified input–output relation and estimate losses based on average currents. Let us assume that the only source of losses is the inductor. The modified expressions for the output voltages of the two basic converters in CCM are given by

$$V_{o,buck} = \frac{DV_i}{1 + R_L/R}, \quad V_{o,boost} = \frac{V_i}{1 - D}\frac{1}{1 + \frac{R_L}{(1-D)^2 R}}. \tag{3.9}$$

In the case of a buck converter the only consequence is the slightly lower output voltage. In the case of a boost converter the output will first increase, but after a certain point it will begin to decrease. The shape of the curve and the exact location of the V_o maximum depends on the ratio of R_L and R. Some examples are illustrated in Figure 3.10. The first conclusion to make is that the maximum output voltage of the boost converter is limited by losses, and the second is that the slope polarity reverses. As a consequence, the negative feedback used to control the output becomes positive for high values of the duty cycle, causing the converter to fail. The way to avoid this is to limit the duty cycle to the region where the derivative of the slope is positive. In practice this means adding some delay circuits and logic gates.

The limitation of the zero-ripple approximation is that it can only be used for the CCM and that the additional losses due to the current ripple are neglected. A different way to approximate losses is to assume that they are relatively small, and that they do not affect the voltage and current waveforms. The above derivations can then be used directly to calculate losses in the switches and the inductor. The approximation is valid as long as the converter efficiency is high, practically this means close to 90% or higher. For both the buck and the boost converter in CCM we can express the losses as the function of the inductor

Figure 3.9: Ratio of output and input voltage for a boost converter with losses.

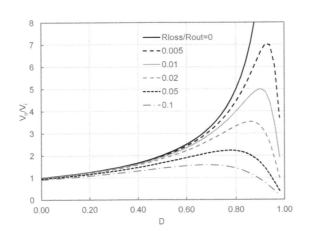

current

$$P_{S1} = DR_{S1}(I_L^2 + \Delta I_L^2/3), \tag{3.10}$$

$$P_{S2} = (1 - D)R_{S2}(I_L^2 + \Delta I_L^2/3), \tag{3.11}$$

$$P_L = R_L(I_L^2 + \Delta I_L^2/3), \tag{3.12}$$

where I_L is the average current through the inductor. In the buck converter case it is equal to the output current $I_L = I_o$, so the losses are directly proportional to the square of the output current. In the boost converter the inductor current is equal to the input current $I_L = I_i$, which is a function of the output current as $I_L = I_o/(1 - D)$, resulting in the same conclusion as for the buck converter.

Unsurprisingly, the power dissipated in the two switches depends on the duty cycle, or in other words the on-time of each switch. Another thing to notice is that it also depends on the ripple amplitude. The larger the ripple, the higher the losses in the switches and the inductor for the same load current. An obvious way to reduce the current ripple is to increase inductance, but this also increases the inductor resistance R_L. The inductance of a coil is proportional to the square of the number of turns. Increasing the number of turns means longer wire, and wire resistance is proportional to its length. In the first order approximation, the inductor resistance should increase proportionally to the square root of inductance. However, the equivalent series resistance of the inductor is not just due to the conductor loss, it is a combination of different effects, and, just to make things more fun, it is usually frequency dependent. The second obvious way to reduce the current ripple is to increase the switching frequency, but this also leads to an increase in switching losses that will be discussed in a bit.

For the DCM the inductor and switch currents are triangular and the losses depend only on the peak current. Again, for both buck and boost converters the expressions are identical and given by

$$P_{S1} = D_1 R_{S1} I_p^2 / 3, \tag{3.13}$$

$$P_{S2} = D_2 R_{S2} I_p^2 / 3, \tag{3.14}$$

$$P_L = (D_1 + D_2) R_L I_p^2 / 3. \tag{3.15}$$

The peak current of the buck converter in DCM can be calculated as $I_P = D_1(V_i - V_o)/f_s L = 2I_o/(D_1 + D_2)$, and D_2 derived directly from the volt-second balance. The peak current expression is slightly simpler for the boost converter $I_P = D_1 V_i / f_s L = 2I_o/D_2$. In both cases the peak current depends on the load and the duty cycle D_1. In general, the conduction losses will always present a fraction of the output power. In that sense, as a very coarse approximation, the overall efficiency should be constant regardless of the load, and it can be maximized by simply minimizing the resistance of the switches. In reality this is not the case because of the other loss mechanisms present in a converter.

The second important factor to consider are the switching losses. Because the resistance of the switches should be small, the switches themselves tend to be large, meaning that they pose a considerable capacitive load for the driving circuit. The power needed to drive the switches is given by

$$P_{switch} = f_S (C_N + C_P) V_H^2. \tag{3.16}$$

Clearly, higher switching frequency results in higher switching losses. The switching frequency indirectly affects the conductive losses. Lower current ripple due to higher switching frequency should somewhat reduce the conductive losses in the switches. In both the buck and the boost converter the switches should be driven by the higher of the two voltages, so for the buck $V_H = V_i$ and for the boost $V_H = V_o$. This is the case in equilibrium, but bear in mind that when the boost converter is starting the output is at 0 and another supply must be used for the switches. For a MOS transistor, the on-resistance and the gate capacitance are given by [3]

$$R_{ON} = \frac{1}{\mu C_{ox} \frac{W}{L} \left(V_G - V_{T0} - \frac{n}{2}(V_D + V_S) \right)}, \tag{3.17}$$

$$C_G = WLC_{ox}. \tag{3.18}$$

Figure 3.10: Losses on a switching MOS transistor.

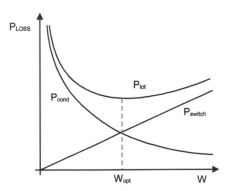

Here, W and L are the width and length of transistor, C_{ox} is the oxide capacitance, μ is the mobility. The key observation is that the resistance decreases proportionally to $1/W$, whereas the capacitance increases proportionally with the W. For a given operating point of the converter, fixed switching frequency and the inductor parameters it is possible to find an optimal size of the transistors to minimize losses. Changing any of the converter parameters changes the optimal width of the transistor. Often, however, there will be more than one point that require a decent efficiency, so finding a global optimum, including the optimal inductor and switching frequency, becomes a complex optimization problem.

Several other loss mechanisms can be identified, but these are usually low compared to conduction and switching losses. Once the main contributors have been optimized you can work on the rest to squeeze out that last percent of efficiency. The first one is the short-circuit current, which may appear if the two switches are on at the same time. This problem is easily solved by using non-overlapping driving pulses, such that there is a very short dead time when neither of the two switches is conducting. Again in this case, one of the parasitic switch diodes will conduct, but as long as the dead time is small compared to the switching period, this loss should be negligible. Another contributor proportional to the switching frequency is the loss due to the parasitic capacitance in the V_{sw} node, equal to $f_s C_{sw} V^2$, where V is the difference in voltages on the node. As mentioned before, in a synchronous converter in DCM, some additional losses will appear if the switch is not ideally opened when the inductor current reaches 0. In that sense, it is better to open the switch slightly early to avoid the discharge of the output capacitor through the inductor. In either case, the dissipation on the parasitic diode won't hurt the efficiency as long as its on time is small compared to the switching period.

In addition to these losses there are a number of effects neglected in the shown formulas. To mention some, the gate capacitance and switch resistance are not constant during the on period, both depend on the gate voltage and drain current of the transistor. Furthermore, the switches are not turned on instantaneously, and there will be a short period during which the switch resistance is significantly higher. None of these effects is devastating on its own, but when all of them are combined they could affect the optimal size of the switches. It is important not to forget that all the formulas shown here are very approximate. Their purpose is to develop an intuition for designing a converter and to provide a solid starting point in the component optimization. The fine tuning must be done through simulation that includes all the effects neglected here.

Another factor should be accounted for when optimizing the converter efficiency. This is the consumption of the regulator, used to maintain the constant output voltage and generate the driving signals for the switches. A standard PWM controller (more on this in the following sections) could easily consume around 100 µA or above. For most people dealing with power electronics, such a number would be under the radar. But, it becomes important when designing a PMU for a low-power SoC. At peak consumption of the SoC, the PWM controller is likely a negligible factor, however in the sleep mode the SoC could consume less than 10 µA, but the supplies must be kept on. It is clear how a DC–DC converter consuming 100 µA is not very useful in such a scenario. In fact in the described case an LDO might be more efficient due to its simplicity, even if the dropout voltage is not very low. An alternative approach is to use a DC–DC converter with a less standard regulator, optimized for low load currents. Similarly to the SoC, the DC–DC converter can also have different modes of operation and adapt to different scenarios, as will be described later in this chapter.

3.3 DC–DC Converter Modeling

3.3.1 Voltage mode control

The DC–DC converters require a feedback loop to control the output voltage. The duty cycle D, that controls the on time of the switch S_1, is often used as a control variable, which is adjusted to keep the output voltage constant. In order to close the loop it is important to understand the dynamic behavior of the DC–DC converters. What we are interested in is how the average value of the output voltage changes as a function of other parameters (Figure 3.11). Once again, we

Figure 3.11: Real and averaged output voltage waveform.

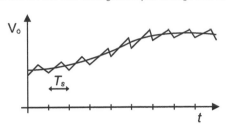

are neglecting the small switching ripple and focusing on what happens to the averaged values of different quantities, and by this we mean values averaged over one switching period[1]. A sketch of the real and the averaged output voltage is shown in Figure 3.12. Through averaging a nonlinear, continuous-time model can be derived and then used to design a system in a way that we are more familiar with. This nonlinear circuit can then be linearized around a desired operating point to obtain a linear model and subsequently transfer functions that can be used to design the control loop. The large signal averaged model can also be used for transient and AC simulations (here the simulator performs the linearization for you), allowing to save considerable time compared to a full simulation of the switching converter. Naturally, the averaged model is only valid for frequencies sufficiently below the switching frequency, and becomes less precise the closer you get to it.

There are different ways to derive the averaged model described in the literature, but the most intuitive is likely the "averaged switch" model. The key observation is that the switch is the only time-varying element in the circuit. The approach is to replace the switches with current and voltage sources that generate the same average currents and voltages as in the original circuit. The time-invariant circuit is then used for further analysis, and the standard perturbation method can be used to linearize it. It is interesting to note that the switch topology is the same for all the basic converters, so the same averaged switch model can be used in any of them without modification (aside from a change in the current direction through the switch S_2 in the buck and boost case).

For the buck in CCM it is sufficient to look at the voltage and current waveforms in Figure 3.4(a). It can easily be concluded that the input current

---

[1]There will be no particular notation to distinguish averaged currents and voltages from actual waveforms, however the distinction should be clear from the context.

Figure 3.12: Buck converter with the enclosed switch element.

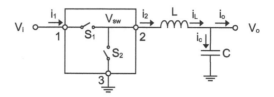

of the switch i_1 is equal to di_2 and that the average output voltage v_1 is dv_2. In DCM the derivation is slightly more complicated, but it is still based on the steady state waveforms from Figure 3.4(b). The following current and voltage relations can be obtained:

$$i_2 = \frac{d_1 + d_2}{d_1}i_1, \tag{3.19}$$

$$v_2 = \frac{d_1}{d_1 + d_2}v_1, \tag{3.20}$$

$$i_2 = \frac{d_1 + d_2}{2}I_P. \tag{3.21}$$

An additional equation is needed to eliminate d_2 from the above expressions. The peak current I_P can be expressed in terms of the switch voltages if we assume that the average inductor voltage is 0. This is only justified in steady state, but the approximation is nevertheless valid at low frequencies (relative to the switching frequency). We can then write:

$$v_{12} = L\frac{I_P}{d_1 T_s} = \frac{2L}{d_1^2 T_s}\frac{d_1}{d_1 + d_2}i_2 = \frac{2L}{d_1^2 T_s}i_1. \tag{3.22}$$

Following the approach from [15], we can define the effective resistance as $R_e(d_1) = \frac{2L}{d_1 T_s}$, and the model is then fully described by the following two equations:

$$v_{12} = R_e(d_1)i_1, \tag{3.23}$$

$$i_2 = i_1 + v_{12}i_1/v_2 = i_1 + R_e(d_1)i_1^2/v_2. \tag{3.24}$$

The current at the second terminal is a sum of two components. The first one is the input current. The second component is a current such that the product of this current and voltage at the switch terminal 2 is constant. In other words it's a constant power source, and the switch output current will be determined by this equation and the load characteristic. The CCM and DCM averaged switch models are given in Figure 3.13.

A slightly different modeling approach can be taken that is quite convenient for simulations [16]. In fact, by looking at Equations (3.19) and (3.20) for DCM,

Figure 3.13: Buck converter averaged switch model in CCM (a) and DCM (b).

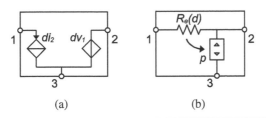

(a)   (b)

Figure 3.14: Buck converter averaged switch model valid for CCM and DCM.

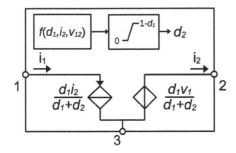

we should notice that, just like the CCM equations, they describe an ideal transformer. Furthermore, they are equivalent if $d_1 + d_2 = 1$, which is a natural limit between the CCM and DCM. Starting again from (3.21), it is possible to calculate d_2 as

$$d_2 = \frac{2Lf_s i_2}{d_1 v_{12}} - d_1. \tag{3.25}$$

The model can be constructed as shown in Figure 3.14, the only additional element required is a clamp for d_2 that limits it to the interval between 0 and $1 - d_1$. For simulation convergence, a soft limiter implemented with diodes (exponential characteristic) is usually a preferred choice than a piece-wise linear clamp, as explained in [16]. The power of this model is that it can be used for large and small signal simulation and that it already includes a transition between CCM and DCM, but it is not the most intuitive one for hand calculation. Interestingly, it can be directly used for a boost converter as well.

The small signal model is obtained by linearization around the operating point, it is a straightforward, but somewhat tedious, process (especially for the DCM), and will not be shown here, but can be found in [15]. The small signal model is then used to derive different transfer functions, but owing to the fact that we're dealing with relatively low currents, the issues that appear elsewhere generally don't show up here. At the converter input we will typically find a battery and a large decoupling capacitor. The input impedance and line-to-output functions are rarely a concern in this case. What is of concern is the input peak current, which shouldn't exceed the battery limits. The large output capacitor will normally be sufficient to absorb any rapid load variations, and hence there is usually no need to worry about the output impedance either. Naturally all of these parameters need to be verified for the final design, but should not present a major challenge in the design process. For the low-power integrated converter, the control-to-output transfer function is mainly of interest, as it is needed for the compensator (controller) design. For both buck and boost converter in CCM, the control-to-output transfer function is given by

$$H_{CCM}(s) = G_0 \frac{1 - \frac{s}{\omega_z}}{1 + \frac{s}{\omega_0 Q} + \frac{s^2}{\omega_0^2}}, \tag{3.26}$$

where different equation parameters are defined as

$$Buck : \ G_0 = V_i, \ \omega_0 = 1/\sqrt{LC}, \ Q = R\sqrt{C/L}, \ \omega_z = \infty, \tag{3.27}$$

$$Boost : \ G_0 = V_o/D', \ \omega_0 = D'/\sqrt{LC}, \ Q = D'R\sqrt{C/L}, \ \omega_z = D'^2 R/L, \tag{3.28}$$

where capital letters are used to denote DC values. In both cases the denominator is a second order function, coming from the LC network. In this case, a resistive load is assumed, but the result can easily be derived for a fixed load current, and added series resistance of the inductor and capacitor. Inclusion of the capacitor ESR will result in added LHP zero in the transfer function. One significant difference is the absence of the RHP zero in the buck converter, simplifying the compensation a bit. Also, the bandwidth of the boost converter is slightly lower due to the D' factor.

The small signal model in the DCM is somewhat more complex, although it results in a simpler control-to-output transfer function. The general expression for basic converters is given by

$$H_{DCM}(s) = \frac{G_0}{1 + \frac{s}{\omega_p}}, \tag{3.29}$$

which is a single pole transfer function. In practice, the compensator is designed to assure a good phase margin for a converter in CCM, while in DCM the converter often ends up stable without any added effort. The parameters for the two converters are given by

$$Buck : \ G_0 = \frac{2V_o}{D} \frac{1 - M}{2 - M}, \ \omega_p = \frac{2 - M}{(1 - M)RC}, \tag{3.30}$$

$$Boost : \ G_0 = \frac{2V_o}{D} \frac{M - 1}{2M - 1}, \ \omega_p = \frac{2M - 1}{(M - 1)RC}. \tag{3.31}$$

Figure 3.15: Voltage mode buck converter with a PWM.

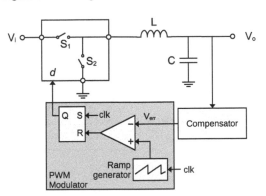

The equations are valid for converters with a resistive load, and without losses. Again, addition of the capacitor ESR results in a new LHP zero. As we mentioned before, in the DCM analysis we assumed a zero average voltage across the inductor. Such an approximation led to the above equations, without this assumption, a second pole appears at higher frequencies [15], however it can usually be neglected as it is close to the switching frequency, and shouldn't have an impact on stability.

So far we have looked at the averaged dynamic behavior of the switch, and as we said the switch is controlled by the duty cycle d. When closing the feedback, the output voltage will be used as an input for the compensator, used to provide a stable overall loop gain. The compensator provides the error voltage V_{err} at its output. The block that is still missing is the block that converts the error voltage into pulses of duty cycle d. This is block is the pulse-width modulator, shown in Figure 3.15. It consists of a saw-tooth waveform generator and a comparator. In each clock cycle the saw-tooth waveform linearly rises from 0 to the maximum value V_M. The comparator output flips to 1 each time the saw-tooth voltage exceeds the error voltage resulting in a rectangular output whose duty cycle is equal to V_{err}/V_M. From the small signal stand point this is simply equivalent to the multiplication by a constant factor $1/V_M$, which is a connection between the input voltage and the output duty cycle.

3.3.2 Current mode control

Current mode (CM) control, or peak current control, is a very practical alternative to the voltage mode control. Instead of controlling the duty cycle of

Figure 3.16: Buck converter current mode switch.

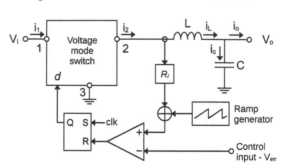

Figure 3.16: Buck converter current mode switch.

the switch directly, a second, fast control loop is added on top. This loop regulates the duty cycle D such that the peak current corresponds to the value defined by the new control input. By doing this, in a very first order approximation, the switch is turned into a current source with a controlled average current. Since this source is in series with the inductor, the inductance no longer contributes to the small signal transfer function, effectively reducing the number of poles and allowing for a relatively simple compensation. A block diagram of the CM controller is given in Figure 3.16.

There is some added complexity to the CM control. Obviously, a current sensor is now needed. For the integrated implementation, this sensor can be implemented by mirroring the switch current, and then measuring the voltage that the mirrored current generates in a known resistor. Addition of a current sensor is a useful feature as it also provides the capability to limit the output current in order to avoid exceeding the switch or inductor limits. One other particularity of the CMC is that the sub-harmonic oscillations appear for duty cycles above 0.5. To prevent them, an artificial current ramp signal can be added to the current sensor ramp [15, 16]. It is possible to derive a stability criterion $m_a > m_2/2$, where m_a is the artificial slope and m_2 is the falling current slope (see Figure 3.17).

The derivation of the CCM model is similar to the VM case, with a modified control input and the addition of the current slope, as shown in Figure 3.17. The previously valid equations for the switch input and output voltages and currents are still valid, $i_1 = di_2$ and $v_2 = dv_1$. Rising and falling current slopes are still related to the input and output voltages as $m_1 = v_{12}/L$ and $m_2 = v_2/L$. To derive the output current expression, we need to use a bit of geometry. Looking at y-values of points a, b and c we get

$$b = a - \frac{m_a d T_s}{R_i}, \quad c = b - \frac{m_2 d' T_s}{2}. \tag{3.32}$$

Figure 3.17: Current-mode CCM inductor current waveform.

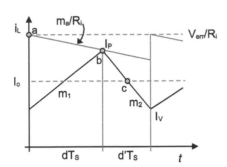

Figure 3.18: Current-mode DCM inductor current waveform.

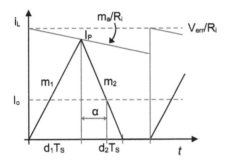

We can then express the slopes in terms of voltages and arrive at the expression for the switch output current

$$i_c = \frac{v_{err}}{R_i} - \frac{m_a}{R_i}dT_s - v_2(1-d)\frac{T_s}{2L} = \frac{v_{err}}{R_i} - I_\mu. \tag{3.33}$$

This expression defines the output current as a function of different parameters, including the duty cycle d, which is no longer a control variable as in the previous case, but can be expressed as $d = v_2/v_1$. These equations model the low frequency behavior. Without going into detail here, the additional high-frequency effects, close to one half of the switching frequency, can be modeled by placing an additional capacitor $C_s = 4/(L\omega_s)$ between terminals 2 and 3 [16, 17]. This capacitor models the delay between a perturbation and the effect on the current slopes, that is delayed by one cycle. The memory effect is not present in the DCM as the current always returns to zero.

We can now focus on the DCM in order to derive a model with a gradual transition between the two modes. Similarly to the VM case the DCM voltage and current relations $v_2 = \frac{d_1}{(d_1+d_2)}v_1$ and $i_1 = \frac{d_1}{(d_1+d_2)}i_2$ are valid, and again reduce to CCM equations for $d_1 + d_2 = 1$. The current slope expressions are equal to the ones in CCM. We can refer to Figure 3.18 for deriving the output current expression. The peak current is given by

$$I_P = \frac{v_{err}}{R_i} - d_1 T_s \frac{m_a}{R_i},\tag{3.34}$$

accounting for the reduction due to the ramp. Knowing that

$$i_2 = (d_1 + d_2)I_P/2 = I_P - \alpha d_2 T_s m_2,\tag{3.35}$$

where α is an unknown parameter, shown in Figure 3.18, we can use the similarity of triangles to obtain

$$\alpha = 1 - \frac{i_2}{I_P} = 1 - \frac{d_1 + d_2}{2}.\tag{3.36}$$

By inserting (3.34) into (3.35), and replacing the current slope m_2 and α, with their expressions yields

$$i_2 = \frac{v_{err}}{R_i} - \frac{m_a}{R_i}d_1 T_s - d_2 T_s \frac{v_2}{L}(1 - \frac{d_1 + d_2}{2}) = \frac{v_{err}}{R_i} - I_\mu.\tag{3.37}$$

Two more equations are needed to derive d_1 and d_2. We can get them from the voltage and current relations as

$$d_1 = d_2 \frac{v_2}{v_{12}},\tag{3.38}$$

$$d_2 = \frac{2Li_2}{d_1 v_{12} T_s} - d_1.\tag{3.39}$$

The above three equations constitute a system of nonlinear equations that describe the CM DCM switch. In principle, by inserting (3.38) into (3.39) it is easy to get an explicit expression for d_2 that doesn't depend on d_1, although this is not needed for a model used in simulations. The switch model is shown in Figure 3.19. The capacitor that models the high-frequency CCM behavior can be added in series with a controlled switch that is closed in CCM and open in DCM (but bear in mind that that this may cause convergence issues in transient simulations).

The switch model can be plugged into the boost converter, the only modification required is inverting the sign of the current i_2. The modification is somewhat intuitive if you notice that the direction of the diode inside the switch changed from buck to boost. An alternative approach is to go through the same derivation for the boost converter to arrive at a slightly nicer model adjusted for the boost switch.

Figure 3.19: Current-mode switch model valid for CCM and DCM.

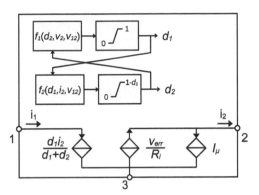

Needless to say, the small signal model derivation based on the above equations is a painful process, but luckily it can be found in the literature [15]. As for the VM case, we will only state the control-to-output transfer functions as they are useful for the controller design. The low-frequency transfer functions for the buck and boost converter are given by [17]

$$Buck : H_{CM,CCM}(s) = \frac{R}{R_i} \frac{1}{1+sRC}, \tag{3.40}$$

$$Boost : H_{CM,CCM}(s) = \frac{R}{2R_i} \frac{1 - \frac{sL}{R}\frac{V_o^2}{V_i^2}}{1 + \frac{sRC}{2}}. \tag{3.41}$$

The simple single pole characteristic is what motivates most designers to use the current mode control. Again, as in the VM, a RHP zero appears in the boost characteristic. Ideal passive components are assumed in the above two formulas, and the presence of ESR in the capacitor will add an LHP zero in both functions. To include the high-frequency effects, the transfer functions should be multiplied by

$$H_{HF} = \frac{1}{1 + s\frac{(1+\frac{m_a}{m_1})(1-D)-0.5}{f_s} + \frac{s^2}{(\pi f_s)^2}}. \tag{3.42}$$

Interestingly, the same factor is used for both buck and boost converter (this part is modeled by the additional capacitor).

In DCM the transfer function remains a single pole one. A single expression can be used for both converters:

$$H_{CM,DCM}(s) = \frac{a(R \parallel r)}{1 + s(R \parallel r)C}, \tag{3.43}$$

where a and r are parameters given by

$$Buck : a = 2\frac{I_o R_i}{V_{err}}, \quad r = R\frac{(1-M)(1+\frac{m_a}{m_1})}{1-2M+\frac{m_a}{m_1}}, \tag{3.44}$$

$$Boost : a = 2\frac{I_o R_i}{V_{err}}, \quad r = R\frac{(M-1)}{M}. \tag{3.45}$$

Although the DCM expressions seem somewhat more complex, only one pole is present in the transfer function. It is relatively easy to design a compensator for the current mode converter in both CCM and DCM, which is the main motive behind the use of this technique.

3.4 DC–DC Converter Control

The previous section dealt with the analysis of the dynamic behavior of DC–DC converters, this is a prerequisite for the controller design. Once the transfer functions of the converter are known, we can provide a suitable compensator to meet the stability requirements. In addition a good compensator should provide a low output impedance and should attenuate any disturbances coming from the input voltage. Now, because here we are dealing with battery powered SoCs, with a peak current in the order of several hundreds of milliamps, this task is somewhat simpler than in the general case. The input battery voltage is not likely to drop out of nowhere, and is relatively clean. One important thing to check is the voltage drop due to finite impedance of the battery during the high load current intervals. This is often solved by placing a large capacitor in parallel with the battery, at the converter input. The large output capacitor is usually sufficient to absorb any load current steps you might encounter, so the output impedance is rarely a concern during the design process. It is nevertheless a good idea to verify and simulate the worst case scenario, to avoid any ugly surprises. The provided averaged switch models can be used to simulate the AC response of the converter and to observe the impact of the compensator and the negative feedback on the input and output impedance and the PSRR.

The primary concern when designing a compensator is the converter stability, quantified by the phase margin (other more sophisticated methods exist, but this one is simple and does the job in most cases). Given the converter transfer function, a compensator should be implemented that provides the required phase margin at the open loop unity-gain frequency. There are different ways to implement compensators. In a typical analog implementation an operational amplifier (OA) or a transconductor is used. In integrated versions the OA is more popular for two reasons. The transconductor is not able to provide a high DC gain, which is usually required for precise load regulation

Figure 3.20: Type 2 compensator schematic (a) and its transfer function (b).

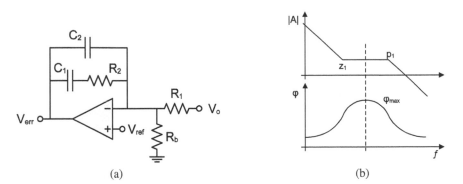

(a)   (b)

and is always good to have in order to reduce the low-frequency output impedance. The second is the variation of the transconductance in PVT corners, which directly affects the loop gain. An operational amplifier in a closed loop configuration provides fairly stable characteristics even in PVT corners.

Two standard compensator implementations using an OA are given in Figures 3.20 and 3.21. These are referred to as the type 2 and type 3 compensators, the name simply reflecting the number of poles in the transfer function. Aside from poles, both of these compensators also have LHP zeros, which provide phase boosting. Because all the converters already have at least one pole in their transfer function, phase boosting is needed for proper stabilization. This is why a type 1 compensator, i.e. a simple integrator, cannot be used here. Note that in both cases the zeros must be located at lower frequencies than poles, otherwise we would see a phase drop. For the type 2 compensator the transfer function is given by

$$\frac{V_{err}}{V_o} = \frac{1 + sC_1 R_2}{sR_1(C_1 + C_2)(1 + s\frac{C_1 C_2}{C_1 + C_2}R2)}. \tag{3.46}$$

When choosing the values of passive components, C_1 and R_2 can be chosen first to set the zero frequency, and then C_2 is chosen to fix the pole frequency. The resistor R_b doesn't affect the transfer function, and is used with R_1 as a resistive divider that the defines the output voltage. The type 2 compensator is often used with the current mode converter, as the single pole provides a sufficient phase boost. The frequency of the phase maximum appears at the geometric mean of the zero and pole frequency $\sqrt{f_z f_p}$, while the maximum itself is determined by the separation between them. They are chosen such as to have the crossover or unity-gain frequency between them, and need to be sufficiently far apart to obtain the desired phase boost. At the same time the pole should

Figure 3.21: Type 3 compensator schematic (a) and its transfer function (b).

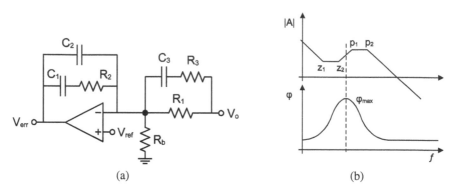

be sufficiently low to limit the high frequency gain. Recall that some peaking might occur in CM converters at one half of the switching frequency; the pole should be placed such that this peak remains below 0 dB to avoid any oscillatory behavior.

Type 3 compensators are used in VM converters in CCM, in order to deal with a double pole transfer function. The type 3 compensator transfer function is given by

$$\frac{V_{err}}{V_o} = \frac{(1 + sC_1R_2)(1 + s(R_1 + R_3)C_3)}{sR_1(C_1 + C_2)(1 + s\frac{C_1C_2}{C_1+C_2}R2)(1 + sC_3R_3)}. \tag{3.47}$$

The same pole and zero are present as in the type 2 compensator, plus an additional pole–zero pair coming from R_3 and C_3. For a VM converter compensation a good approach is to make sure the LC resonant frequency is sufficiently away from the crossover frequency, to avoid oscillations due to load perturbations [18]. The two zeros can be placed close to the resonant frequency, while the crossover frequency should be at least 3–5 times higher than the resonant frequency. The two poles should be placed after the resonant frequency in order to maximize the phase margin. If the zero related to the capacitor ESR appears below the crossover frequency one of the poles can be used neutralize it. Poles should be placed below or close to one half of the switching frequency. An example compensated buck converter transfer function is shown in Figure 3.22. In very simple terms, the aim is to position the crossover frequency in a way that guarantees the minimum required phase margin. More systematic ways to approach compensation are given in [18]. One other thing to have in mind is that the on-chip resistors and capacitors may vary up to 10–20%, so stability should be assured for the extreme values of these components as

Figure 3.22: Buck converter VM transfer function and compensated loop gain.

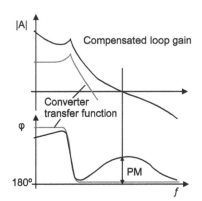

well. In addition, as the converter transfer functions depend on the DC voltages and currents, it is important to verify stability for the entire operating range.

If you have paid attention so far you may have noticed that quite a few blocks are needed to implement a controller for a buck or boost converter. A full controller block diagram is shown in Figure 3.23. The current sensor is needed for the CM control; however, even for the VM controller a current limiter is recommended in order to prevent high current that could damage the switches or the inductor, so the difference in complexity is not that big. Before we have a look at each of the blocks shown in the block diagram, it's worth noticing that there is also a number of blocks that are assumed to be present in the system. First of all, fixed switching frequency converters, such as this one, require a clock signal. Such a signal can be generated by a free running ring or an RC oscillator as the frequency stability is generally not a concern. All the comparators, OTAs and OA require a bias current, which is typically provided by a PTAT source. Finally, a temperature independent reference voltage needs to be provided by an on-chip bandgap reference, or by some external circuit. The bandgap reference can also be used to generate the temperature independent reference current, which is useful for the ramp generation circuit. These circuits will be discussed in the next chapter.

The PWM logic block in its most basic form is an SR latch. On every rising edge of the clock the latch is set, activating the PMOS, and then reset when the PWM comparator changes state, activating the NMOS switch (switches are used in a reverse order in the boost converter). A few additional gates are used to stop the converter and reactivate it based on signals from other comparators.

Figure 3.23: Buck converter current mode controller diagram.

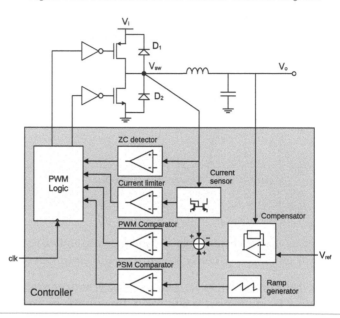

This block is also in charge of generating the non-overlapping pulses for the two switches to avoid the short-circuit current.

The compensator schematics are shown in this section, but a few comments are in order about the operational amplifier design, as it is the crucial block of the controller. In general, a standard two-stage OTA could be used and one important thing to have in mind is the output current capacity. This is because the feedback capacitors tend to be large, as the poles and zeros are located close to the filter resonant frequency. To some extent, larger resistors can be used in order to reduce the capacitor sizes and maintain the RC constants. Still, in some cases a class AB output stage might be necessary in order to avoid slewing behavior of the amplifier. Different solutions can be found in the literature [1], and a simple amplifier with a current mirror is shown here in Figure 3.24. The output stage uses differential inputs, provided directly by the input differential amplifier. A resistive common-mode feedback is used in the shown example, through the resistors R_f. Depending on the gain requirements, additional stages can be placed, although two stages are often sufficient. One additional thing to bear in mind is that the OA gain-bandwidth needs to be above the compensator crossover frequency, to avoid affecting its transfer function set by the passive components. We mentioned in the previous chapter that

Figure 3.24: Compensator operational amplifier with a class AB output.

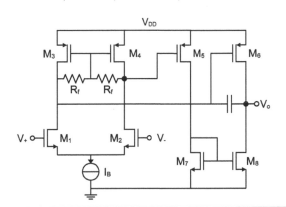

GBW trades with the current consumption, so the minimum GBW compensator requirement will dictate the consumption of the OA.

A simple ramp generator is shown in Figure 3.25. A voltage ramp is first generated by injecting a constant current into a capacitor $V_{ramp} = I_B t / C$. The input clock is used to reset the voltage at each rising edge of the clock signal. Note that the standard clock signal with a 50% duty cycle cannot be used directly, instead, a delay circuit and an "and" gate can be used to implement a rising edge differentiator that will provide narrow reset pulses. In this type of implementation, a real current source I_B could be affected by the V_{ramp}, causing it to deviate from an ideal linear waveform. However, this nonlinearity is not a concern as the converter is a closed loop system, and the feedback will simply provide the error voltage that results in the correct duty cycle of the converter switch. In the second step the error voltage is converted into current using an OTA and a resistor; this is only needed for a CM converter. This current can then simply be added to the sensed inductor current using current mirrors. The sum of the two currents is injected into the resistor R_i (this is one of the averaged switch parameters that appears in the converter transfer function) in order to generate the voltage input for the PWM comparator. The second input of the PWM comparator is the error voltage. The resistor, capacitor and the current I_B can be used to tune the slope of the ramp to the desired value that depends on the inductor current slope and the expected operating point.

As we mentioned already, in order to implement current mode control or a current limiter, it is necessary to sense the inductor current. Actually, for the buck converter it suffices to monitor the current through a PMOS when it's conducting. In the boost converter the inductor current increases while the

Figure 3.25: Simple clocked ramp generator schematic.

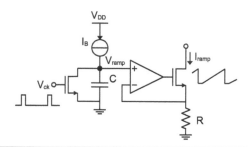

NMOS is conducting, so here you'd monitor the current of the NMOS switch. In both cases the same principle is used, the switch current is mirrored using a large ratio current mirror. This is shown in Figure 3.26 where M_{sen} is K times smaller than the main PMOS switch M_{PSW} [19]. The sensed current can be injected directly into a resistor R_i together with the ramp current. Transistor M_1 must be sufficiently large so that the voltage drop across it is negligible. The error amplifier then ensures the drains of the sensing and switching transistor are at the same potential in order to provide a precise copy of the output current. The transistor M_2 provides a bias point when M_1 and M_{PSW} are off. When the main switch is turned on and the current sensor activated, some time will be needed for the sensor to start tracking the current accurately. In addition, some overshoot may appear resulting in non-monotonic behavior of the current sensor. To deal with this it is possible to introduce a minimum limit for the converter duty cycle, for example, over the first 20% of the switching period the controller waveforms are ignored, giving all the circuits sufficient time to settle avoiding any glitches due to switching. Note that it is also possible to duty cycle the current sensor together with the main switch as it only needs to provide a copy of the current when the switch is conducting, allowing to reduce the controller consumption.

In the integrated version of a DC–DC converter the diode is replaced by a MOS switch. To emulate the diode behavior and avoid draining the load capacitor through the NMOS, a zero-crossing (ZC) detector must be used. As the name suggests, the detector should react when the inductor current reaches 0. In this case there is no need for a current sensor, instead it is possible to detect zero crossing of the NMOS switch voltage, that is V_{sw}. Any standard comparator can be used, although it is possible to exploit the fact that the input voltage is compared to 0. The implementation from [20] is shown in Figure 3.27. In addition to the standard differential pair input, the source of M_4 is also used as input, allowing to speed up the comparator. The delay of the comparator is

Figure 3.26: Current sensor schematic.

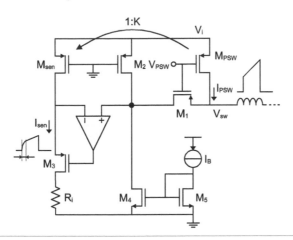

important as it translates into the current offset that causes losses in the switch and the parasitic body diode. As long as the delay is small this loss shouldn't affect the efficiency significantly. One way to improve it is to add an offset to the comparator, so that it is triggered a bit in advance in order to turn off the switch exactly when the inductor current is 0. Unfortunately, given that the comparator delay changes with the PVT variations and that different operating points require different delays, it is impossible to cover all the cases with a single offset value. It is possible to implement a calibration loop that tracks changes and adjusts the comparator offset, but the added complexity is only justified if this is the factor limiting efficiency, which is usually not the case. Just like the current sensor, the ZC detector can be duty cycled together with the NMOS switch of the buck converter (PMOS in the boost converter).

Three more comparators are left in the controller. In principle, any standard design based on an input differential pair can be used. The important parameters to have in mind are the input voltage range and the delay. The most important one is the PWM comparator. Its role is to flip the state when the sum of the ramp and the current sensor voltage exceed the error voltage (here the error voltage changes slowly with respect to the switching period). Typically it must be able to handle the rail-to-rail input voltages and must react fast. Although the delay will be compensated by the overall negative feedback loop, the comparator delay must be a small fraction of the switching period to avoid limiting of the achievable duty cycle. The comparator delay depends heavily on its consumption, so this one is likely to be the most power hungry comparator in the system. The current-limiting comparator is used to open the switch if the

Figure 3.27: Zero crossing detector schematic.

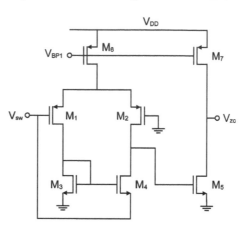

inductor current reaches the maximum value. Again, the delay is important here to avoid exceeding the current limit, but the problem can be avoided by using a small offset and triggering the comparator in advance. Unlike with the zero crossing detector, there is no penalty for efficiency here.

In cases when the load current is low, the converter might have difficulty in providing the desired voltage level. This may be the case because there is minimum duty cycle which determines the minimum on-time of the PMOS switch in the buck converter, consequently setting the minimum current the converter injects into the load. If the load current is below this value, the output voltage begins to rise. To avoid this behavior the pulse skip mode (PSM) is introduced. In essence, when such behavior is detected the converter is stopped and waits until the output capacitor is drained sufficiently by the load current before restarting normal operation. The very low value of the error voltage can be used to indicate that the converter should switch to PSM and, once it recovers, the normal operation can continue. Because the output capacitor is large and the output voltage rises slowly, the PSM comparator doesn't need to be particularly fast. A comparator with hysteresis should be used to avoid glitches when the error voltage is close to the PSM limit. The PSM reference voltage should be set relative to the ramp and the current sensor outputs, allowing the duty-cycle to be limited to the desired range. Since switching to the PSM indicates a low load condition, it can be used to switch to a different mode of control, one that is more efficient in such a scenario. This is the topic of the next section.

Figure 3.28: Typical efficiency curves for the PFM and PWM controller as a function of load current.

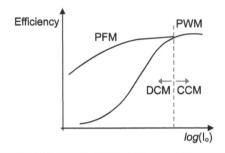

3.5 Low Power Techniques

The standard fixed frequency voltage or current mode control is useful when the load current is significantly larger than the controller current. Battery powered SoCs only consume high current for short periods of time, and spend most of the time in a low power mode (sleep mode, retention mode) where their consumption is reduced to a minimum. Often only a 32 kHz clock and a small portion of memory keeping the CPU context are left on, with a total consumption below 1–10 µA. It is important to provide good efficiency in this mode of operation as it will often be a limiting factor to the battery lifetime.

As shown in the previous section, the standard fixed switching frequency controllers tend to get quite complex and may consume well above 100 µA of current, which limits their efficiency at very light loads. In such cases simpler controllers are preferred, often using some form of pulse frequency modulation (PFM) where the switching frequency is proportional to the load current. This feature allows for a good efficiency over a large range of load currents, which is illustrated in Figure 3.28 (note that the curves are plotted for a logarithmic scale of the load current).

A class of PFM regulators that is often used for good light load efficiency is the ripple regulators [21]. The main reason they are useful is the simple architecture that doesn't require an error amplifier. In addition, their reaction time to load or line perturbations is quite fast (several switching cycles), but they also have a number of disadvantages, which is why a PWM is more often used at higher loads. First of all, varying switching frequency, that in some implementations depends on the capacitor parasitics, can be a problem (mainly if this frequency overlaps with the frequency of a sensitive signal). Since the

Figure 3.29: Buck converter with a hysteretic controller.

Figure 3.29: Buck converter with a hysteretic controller.

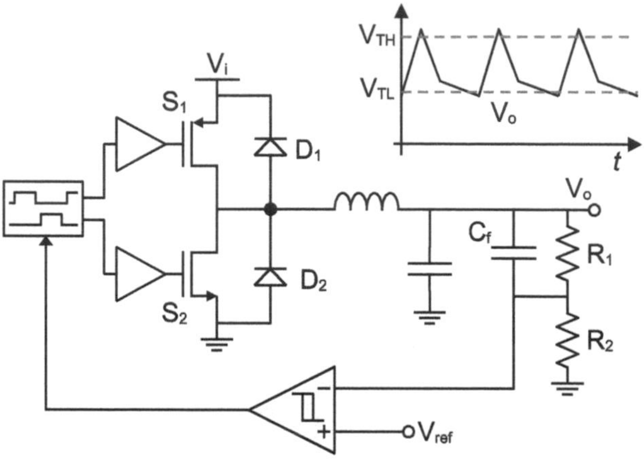

output voltage variation is often quite small, the regulator will be sensitive to noise and coupling from other signals. DC regulation is not as precise as with the PWM regulator and the average output voltage, as well as the voltage ripple, may depend on the load current. This is not necessarily a big issue if an LDO is placed between the converter and the load.

A typical example of a ripple based control is the hysteretic regulator shown in Figure 3.29. In a classical buck implementation with a diode, a comparator with hysteresis is sufficient to provide output voltage regulation. In a synchronous converter the diode emulation can be done using a zero-crossing detector (not shown in the figure). You might think that this is a significant overhead, but remember that it only needs to be on while the NMOS switch is conducting. In DCM, at very light load, this is likely to be a very small fraction of the switching period which makes the impact on the efficiency negligible. It is clear that the current and voltage ripple are highly dependent on the capacitor parasitics, mainly the ESR, but also the capacitance. The delay of the comparator, together with the delay of the logic circuits that drive the switches, also have an impact on the switching frequency and the output ripple. The capacitor C_f is often used in a hysteretic regulator to short the resistor R_1 for the AC signal, meaning that the entire output ripple appears directly at the comparator input, lowering the noise sensitivity.

It is possible to show that the switching period in DCM is proportional to the load current. The approximate on time of the PMOS switch can be calculated as

$$T_P \approx \frac{1}{2}\sqrt{4V_H(1 + \frac{R_1}{R_2})\frac{LC}{V_i - V_o} + R_C^2 C^2} - \frac{1}{2}R_C C, \qquad (3.48)$$

where $V_H = V_{TH} - V_{TL}$ is the comparator hysteresis. Here we assumed that the only parasitic component on the output capacitor is the ESR, denoted as R_C. We also assumed that the output current is much lower than the peak current $I_P = (V_i - V_o)T_P/L$, that the delay of the comparator is negligible and that the output ripple is small compared to the output voltage. For the DCM, the on time of the NMOS switch is simply $T_N = I_P L/V_o$, which is a familiar expression from the previous DCM analysis. Finally, the off time can be calculated using the charge balance on the capacitor, meaning the accumulated charge during $T_P + T_N$ is slowly drained by I_o during T_{off}. The off time is hence given by

$$T_{off} \approx \frac{I_P}{I_o} \frac{T_P + T_N}{2}, \tag{3.49}$$

where we again assumed that $I_P \gg I_o$, which is a valid assumption at light load. The switching frequency is then given by

$$f_{sw} = (T_P + T_N + T_{off})^{-1} \approx \frac{2I_o}{I_P(T_P + T_N)}, \tag{3.50}$$

which is proportional to the load current. As for the losses, the conduction losses will be a fraction of the output power. The switching losses become practically negligible at low switching frequencies. The remaining component is the consumption of the comparator, which becomes dominant for a very low output current.

Hysteretic comparators provide the good low load efficiency that we need, but the dependency on parasitic components makes them unpopular. Similar characteristics can be obtained using a slightly different approach, shown in Figure 3.30. In this implementation the comparator is only detecting the low value of the output voltage (valley). When this critical value is reached, the switches will be activated, charge the capacitor and then let it discharge until it reaches the critical value again. A constant on time for the PMOS switch is used, which determines the amount of charge injected into the capacitor in each cycle, and consequently the current and the voltage ripple. A timer will be required to set the on time, and a bit like the ZCD in the previous example, it will only be used when the PMOS is conducting, making its consumption negligible when averaged over the entire switching period. In some implementations a timer is also used for the on time control of the NMOS switch. The interval is calibrated on the fly, increasing its duration if the current through the inductor is positive when the NMOS switch opens, and decreasing it in the opposite case. The current direction can simply be determined by the sign of the voltage V_{sw}, right after the switch is open. It is important to sample the voltage right after opening the switch, to avoid the ringing when one of the parasitic diodes stops conducting. Alternatively, it is also possible to measure the voltage across the inductor and use it for the calibration loop. For the constant on time regulator the switching frequency is given by

$$f_{sw} = \frac{2LI_oV_o}{T_P^2(V_i - V_o)V_i},$$
(3.51)

which is again proportional to the load current. Clearly, since the minimum (valley) of the output voltage is regulated, the average value will change with the load current, which is one of the downsides of this approach, although if an LDO is placed after the DC–DC converter to regulate the supply, the issue practically vanishes. Just like in the hysteretic converter, the only block that needs to be on all the time is the comparator. Fortunately, this comparator doesn't need to react very quickly, allowing its consumption to be reduced. Because the output voltage changes very slowly if the output current is very low, a clocked comparator can be used to sample its value in regular intervals. Such an approach allows to truly minimize the controller consumption since clocked comparators do not consume power continuously. They will only draw a small amount of current when triggered by the clock signal. Also, in most systems, a low frequency clock is maintained even in sleep mode, so there is no need to implement a separate oscillator. A typical example of the clocked comparator is the strongARM latch [22].

Different variations using a similar approach can be used. One example would be a valley voltage regulator with peak current control, where the timer is replaced by a current sensor. This allows precise control of the peak current, independent of the passive components. In a different approach, only the PMOS switch is used in the low power mode [23]. The NMOS control circuits are put to sleep, and the parasitic diode conducts the inductor current when the PMOS opens. Although the diode losses are generally higher than the NMOS losses, simplifying the control circuits benefits the overall efficiency at very low load current.

Another alternative approach is to add a valley voltage detector to an existing PWM controller to implement a low power mode. Whenever low load conditions are detected, the PWM controller is switched off to conserve power, leaving the comparator to monitor the output voltage. Each time the output voltage reaches the threshold, the PWM controller is turned on for a fixed period of time in order to charge the capacitor. As opposed to the standard PWM mode, a fixed current pulse is more commonly used by overriding the error amplifier and imposing a constant error voltage. Compared to the constant on time regulator, the peak inductor current here is lower, which effectively reduces the resistive losses in the switches and the inductor. Provided that the PWM controller itself doesn't consume too much, this burst mode approach can be more efficient than the simple constant on time approach (Figure 3.31).

Figure 3.30: Buck converter with valley voltage detector and constant on-time.

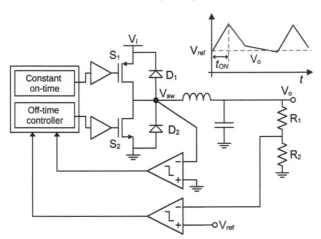

Figure 3.31: Buck converter with a burst mode controller and valley voltage detector.

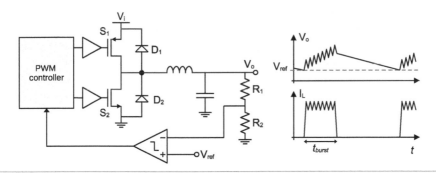

So far we've mentioned several different ways to control the DC–DC converter at different loads, but we never mentioned how and when to switch from one to another. Typically an SoC will have several modes of operation that differ in consumption. Since these modes are controlled by software, it is possible to manually switch between different control modes, given that at least approximate load current is known in advance. If automatic mode switching is required, it can be implemented in different ways. One simple approach is to use low power mode for the DCM and PWM control for the CCM, as the CCM already indicates that a certain load current is exceeded. Both the hysteretic and the constant on time controller can be used in the CCM mode, but special care must be taken to address stability concerns. Some guidelines can be found in [21]. In spite of their good efficiency over a very large current range, they are still not as popular as the standard PWM converter, mainly due to characteristics dependent on the parasitics that are difficult to predict. For a standard PWM converter, PSM detection can be used to indicate low load current and trigger switching to the low power mode. It is then possible to switch back once the load exceeds the current capacity of the low power mode. In other words, the transition occurs when the controller is not able to maintain the output voltage above the defined threshold after the burst.

In the context of battery powered systems, boost converters are commonly used as a part of an energy harvester and a battery charger. The harvester is generally used to extend the battery life, and very rarely will it supply power directly to the system without some form of energy storage. The type of energy source used and the available power will greatly impact the controller of the boost converter. A standard PWM controller is only useful if sufficient input power is available to make the whole converter efficient. Provided that this is the case, it is a preferred choice for its good output regulation and well defined switching frequency. An example use case is a dedicated wireless power transfer system, capable of delivering power in the order of 100 mW to the system. Solar cells are another commonly used energy source. Depending on their size, the available power covers quite a wide power range, but it is typically in the order of 10 mW for a surface of a few cm^2. Other sources, such as thermo-electric generators or mechanical energy harvesters provide fairly low power and a low output voltage, making the power extraction more challenging. Since in all the cases the actual power that can be harvested depends on the environmental conditions and can be very low, a low power controller is almost a mandatory feature in a boost converter.

A hysteretic controller cannot be used in a boost converter; such an approach is limited to converters where the inductor is directly connected to the output [21]. The valley voltage controller with a constant on time can be directly applied to the boost converter. The concept is shown in Figure 3.32. The converter is simply turned on every time the output voltage drops below the threshold, making the switching frequency proportional to the output current. All the variations shown for the buck converter can be directly applied for the boost, with a few minor differences. The peak current is now measured across the NMOS (S_1) switch, so the circuit from Figure 3.26 would be "flipped" and use the NMOS current mirror. For the PMOS (S_2) switch control, a ZCD now compares the input voltage of the PMOS to the converter output voltage. A timer with a ZCD-based calibration loop is often used in place of a direct control with a ZC comparator [24, 25].

Figure 3.32: Boost converter with a valley voltage detector and constant on time control.

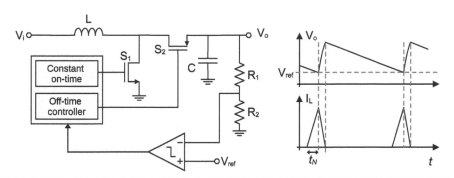

Figure 3.33: Boost converter cold start controller.

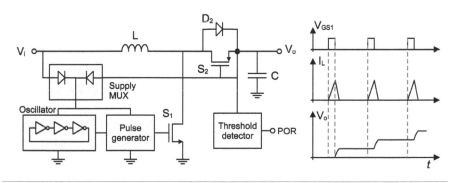

Aside from the output voltage control, some form of input control is required in energy harvesting systems. This is because each source has the optimal bias conditions that allow maximum power extraction. The difficulty is that this point moves with the environmental conditions, and it should be tracked in order to get the maximum out of the harvester. This concept is called maximum power point tracking (MPPT). The way to implement it depends on the source; in some cases it might be sufficient to monitor the output voltage, and to allow the boost converter to draw current only when it is higher than a certain value. In a more complex case both the voltage and the current of the source are monitored and an algorithm searches for the point in which the highest output power is available.

In most systems, the input of the boost converter is connected to the harvester, while the output is used to supply the battery charger. Most batteries, even when almost fully discharged will provide some voltage at their output allowing the converter to use it as the controller supply when starting (a zero output voltage can be a pretty good indicator that the battery is completely dead). In some cases, however the boost converter may be required to start directly from the harvested voltage, which may be fairly low (e.g. below 0.5 V). This feature is known as the "cold start". For example, this may be required because the battery monitoring sub-system must be turned on before drawing the current to avoid damaging the battery. The need for a cold start is generally related to the energy storage element used in the system.

The challenge with the cold start controller is to make it work from a very low supply voltage. The tendency is, naturally, to use the simplest design possible. Fortunately, it is sufficient to control only the NMOS switch, while the parasitic diode of the PMOS can be used to charge the output capacitor. In principle, PMOS control can be added and if the minimum expected input voltage is sufficiently high, this may be a better approach. For the cold start there is no voltage regulation needed, the converter should simply pump the current until the desired output voltage is reached. It is therefore enough to implement a low-voltage oscillator that will provide control pulses for the NMOS switch, as shown in Figure 3.33. A low-voltage oscillator can be implemented as a ring oscillator, which is nothing more than a chain of odd number of CMOS inverters in a loop. This is possible owing to the property of the CMOS logic to operate well below the threshold voltage of the transistor. For a ring oscillator with N stages each inverter produces a 50% duty-cycle signal, with a relative phase shift of $2\pi/N$. One convenient way to generate the desired pulse width for the NMOS driving signal is to combine different phases using a logic "AND" gate. In general, a ring oscillator is quite sensitive to supply variations, which will impact the behavior of the converter; however, this is not of concern for the cold start mode as long the output capacitor can be charged. One other concern in the cold start mode is the resistance of the NMOS switch in the on state, which will be relatively high if the gate voltage is low. A possible way to address this is to add a charge pump to boost the supply for the NMOS driving circuit [24]. The charge pump requires its own oscillator and some logic, but can easily improve the efficiency of the boost converter and speed up the cold start.

The charge pump is only needed while the boost converter output is very low. As soon as the output voltage exceeds the charge pump voltage (or the input supply), it can be used to supply the control blocks. A voltage multiplexer, whose schematic is shown in Figure 3.34, can be used to select the higher of the two supplies. A comparator is used to compare the two supplies and provide control signals for the two PMOS switches. Note that the gate of the non-conducting PMOS is always connected to the higher of the two supplies, ensuring that the transistor is properly switched off and that the leakage is minimized.

Figure 3.34: Supply multiplexer schematic.

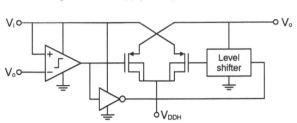

One final thing to note in the cold start mode is the threshold detector that provides the POR signal (power-on reset). Once the output voltage exceeds the threshold this signal will indicate that the supplied circuits can be turned on and that the boost converter can switch to normal mode of operation.

3.6 Capacitive Converters

Capacitive or switched capacitor (SC) converters use switches and capacitors to perform voltage conversion. The advantage, compared to the standard DC–DC converters discussed so far, is the absence of a large inductor, which should allow for full integration. However, in order to supply any substantial current, the capacitive converters require large capacitors, which again results in the need to use off-chip components. They do however find practical use in cases when a small, low-consuming part of the circuit requires a specific voltage. One such example is the error amplifier of an LDO with an NMOS output transistor, that requires a supply voltage above the LDO supply (as seen in the previous chapter). A second typical example in deep sub-nanometer technology nodes is the bandgap reference. This is one of the rare examples of a circuit that still needs BJTs. Since the BJT threshold doesn't scale like the MOS transistor threshold, a standard bandgap reference typically requires a voltage slightly higher than the core supply.

The capacitive converters consist of one or more "fly" capacitors, which are periodically charged and discharged, and a big decoupling capacitor at the output. In some configurations the output decoupling capacitor is not needed, but it helps in reducing the output ripple. We could gain some intuition into the behavior of SC converters if we go back to Figure 3.2, which is a simple voltage doubler. In the first phase ϕ_1 the capacitor C_{FLY} is charged to V_i. In the second phase the capacitor voltage is added to the input voltage and connected to the output. In each clock cycle a part of its charge will be transferred to C_{dec}.

Figure 3.35: Model of a capacitive converter.

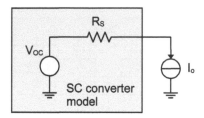

If no load is present, the output voltage will eventually reach $2V_i$. In a general case this output voltage is determined by the configuration of fly capacitors. The input–output ratio can easily be calculated by assuming that the voltage on each capacitor is constant in both phases, which is the case since there won't be any charge transfer in the steady state. If a load current is present the output voltage will drop below its open-circuit value. This voltage drop is proportional to the load current meaning that the capacitive converter can be modeled by a voltage source with a finite output resistance R_S, as shown in Figure 3.35.

To calculate the output resistance, we can assume that the decoupling capacitor is much larger than the fly capacitor, and that in the second phase all of the charge from the fly capacitor is transferred to the output, while the output voltage remains constant. Let us assume also that the output voltage is $V_{OC} - \Delta V = 2V_i - \Delta V$. Each time the fly capacitor is connected to the input source it is charged from $V_i - \Delta V$ to V_i, where the amount of charge that flows into the capacitor is $Q = \Delta V C_{FLY}$. Since we are interested in the steady state, this charge is equal to the charge drained by the load current over one clock cycle $Q_L = I_o T_{sw}$. By using these two equations we get $R_S = \Delta V/I_o = 1/(f_{sw} C_{FLY})$. The output resistance is inversely proportional to the capacitance of the fly capacitor and the switching frequency, and although the calculation was done using a very simple example, this conclusion is valid for all capacitive converters. This implies that in the general case the output voltage can be regulated to some extent by modulating the switching frequency. The main power losses come from charging and discharging the fly capacitors and switch drivers. The total losses in the fly capacitors are equal to $R_S I_o^2$. Typically for a constant switching frequency and capacitor size there is an optimal load current that results in the maximum efficiency. The optimal load current (or load resistor) is a function of the SC converter parameters.

A way to calculate the output resistance in a general case is given in [26]. The technique will be demonstrated on the example of a 3-to-1 ladder converter shown in Figure 3.36, where all the capacitors in the circuit are equal. For

Figure 3.36: Ladder 3-to-1 capacitive converter.

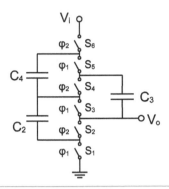

this calculation the assumption is that the output voltage is constant. In other words, we will place an ideal voltage source at the output and calculate the average current, which is the total charge flow into the source divided by the clock period. In order to calculate the output resistance it is necessary to first calculate the charge flows into each capacitor in both phases. The two schematics that correspond to the two phases of the 3-to-1 converter are shown in Figure 3.37. For the calculation of charge flows we can use the fact that the sum of all charge flows into a node is 0 (equivalent of the Kirchoff's current law), and the fact that for all fly capacitors the total charge flow over one clock period must be zero. This translates into $q_i^{(1)} = -q_i^{(2)}$, where the exponent indicates the phase. The charge flows in the example circuit are given by

$$\begin{bmatrix} q_o^{(1)} & q_2^{(1)} & q_3^{(1)} & q_4^{(1)} \end{bmatrix} = q_o \begin{bmatrix} \dfrac{2}{3} & -\dfrac{2}{3} & \dfrac{1}{3} & -\dfrac{1}{3} \end{bmatrix} \tag{3.52}$$

$$\begin{bmatrix} q_o^{(2)} & q_2^{(2)} & q_3^{(2)} & q_4^{(2)} \end{bmatrix} = q_o \begin{bmatrix} \dfrac{1}{3} & \dfrac{2}{3} & -\dfrac{1}{3} & \dfrac{1}{3} \end{bmatrix}, \tag{3.53}$$

where q_o denotes the total charge that flows into the output source during one clock period. These charge flows can then be used to calculate the output impedance based on the following formula

$$R_S = \sum_{i=1}^{N} \left(\frac{q_i}{q_o} \right)^2 \frac{1}{f_{sw} C_i}, \tag{3.54}$$

where N is the number of fly capacitors. In this example we get $R_S = 2/(3 f_{sw} C)$. The formula is derived using Tellegen's theorem, which states that the sum of powers delivered through each branch of a network is 0 (essentially a power conservation law). This formula is valid under the assumption that all the capacitors are fully charged in each phase, which means that the duration of

Figure 3.37: Phases of the ladder 3-to-1 capacitive converter.

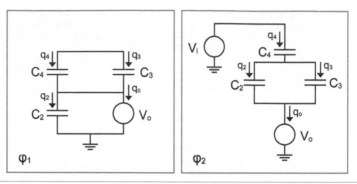

each phase is sufficiently larger than the time constant determined by the capacitance and the switch resistance. This is known as the slow switching limit (SSL). As long as the assumption is valid, the output impedance will be inversely proportional to the switching frequency.

The other asymptotic limit is the fast switching limit (FSL). Because the switching period is now small compared to the RC time constants, the capacitors do not reach equilibrium. In the limit, the capacitor voltages can be considered constant, and all the losses are coming from the on-resistance of the switches. Practically, a nearly linear current will be flowing through the switches. To calculate the converter series resistance, again the same charge flows are used, but now we need to find the amount of charge that flows through each switch when it is conducting. In this case we have

$$[q_{s1} \ \cdots \ q_{s2}] = q_o \left[-\frac{2}{3} \ -\frac{2}{3} \ \frac{1}{3} \ \frac{1}{3} \ \frac{1}{3} \ \frac{1}{3} \right]. \tag{3.55}$$

A single vector is sufficient here as each switch only conducts during one phase. The output resistance is then given by

$$R_S = 2 \sum_{i=1}^{M} \left(\frac{q_{si}}{q_o} \right)^2 R_i, \tag{3.56}$$

where M is the number of switches and R_i is the resistance of the ith switch. It is important to mention that this formula assumes a 50% duty cycle of the clock, i.e. the same duration of both phases, which is often the case in practice. Notice also that the output resistance no longer depends on the frequency. The calculated value is the lowest achievable output impedance of the given converter. Very high switching frequency minimizes resistive losses in the switches, but at the same time increases the switching losses in drivers

Figure 3.38: Schematic of the series-parallel (a) and Cockroft–Walton (b) SC converter

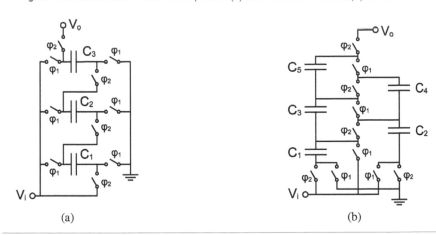

(a)   (b)

Figure 3.39: Output resistance of the three converter types.

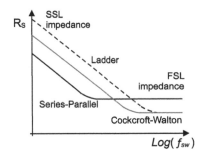

and clock distribution circuits, so it may not necessarily lead to the optimal efficiency. Naturally, the optimal point also depends on the load current as well as the parasitic capacitances of the switches and fly capacitors.

Many different SC converter topologies exist [27], that provide different characteristics. Figure 3.38 shows the series-parallel and the Cockroft–Walton converters. In the figure they are shown as upconverters, although every capacitive converter can perform both roles if the input and output are exchanged. Different topologies provide different output resistances, and interestingly, topologies that provide better SSL output resistance often exhibit worse FSL resistance. Typical curves of the output resistance for the three converters presented here are shown in Figure 3.39.

Figure 3.40: Interleaving of multiple SC converters and the effect on the output voltage

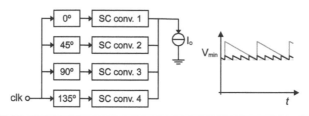

Aside from the output resistance there are other things to take into consideration. For example, different topologies provide different input–output ratios, and require different number of switches and capacitors. The bounds on achievable ratios, and bounds on the minimum number of switches and capacitors are given in [28]. Furthermore, one other thing that is of concern are the switch requirements, as voltages that appear across the switches and needed gate drive depend heavily on the chosen topology and may impact the performance of the converter. As mentioned before, the input–output ratio is fixed for an SC converter. Output voltage can be regulated by regulating the switching frequency; however, this only provides a narrow range of output voltages at a good efficiency. Lowering the output voltage (compared to the open-circuit voltage) increases the voltage drop, which increases the losses in the converter. The "gear-shifting" techniques [29] have been proposed in order to increase the voltage range while maintaining high efficiency. The idea is to make the converter topology configurable by adding more switches, allowing different ideal input–output ratios. The ratio that provides the closest value above the required output voltage is then selected, allowing the operating point to be maintained close to its optimal value for efficiency. A small downside is that this reconfigurability requires additional switches that contribute to slightly increased losses compared to a fixed topology. Another interesting technique is the interleaving, illustrated in Figure 3.40. Instead of using a single converter, the converter can be partitioned into a number of smaller cells that are controlled by different clock phases. This technique primarily allows to reduce the output ripple and use a smaller decoupling capacitor. In some ways, interleaving is similar to increasing the switching frequency, however note that partitioning doesn't change the output minimum voltage, whereas it would shift up if the frequency changed. For a given output voltage, interleaving allows a small efficiency gain compared to a standard SC converter approach without increasing the total capacitor size.

As we've already said, there is a vast number of different SC topologies and techniques reported in academic literature, but in practice they are most often

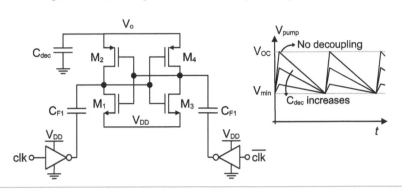

used to provide an auxiliary supply for specific blocks. One example that is often used is the simple voltage doubler shown in Figure 3.41 [30]. Looking at the topology, you will notice that this is in fact an interleaved version of the circuit from Figure 3.2, but here the interleaving can be implemented for free, as the 180° shift is simply an inverted clock signal. Fly capacitors are also used to bootstrap the driving voltage of the four MOS switches to assure low on-resistance, and proper off state. Even though the output voltage is doubled, there is no need to use special transistors, since each transistor only sees a maximum of V_{DD} from drain or gate to source. Latch configuration of the output transistors also regenerates the driving voltage to assure full gate swing. Since in each half-period of the clock at least one capacitor is connected to the output, the decoupling capacitor is not necessary, although it reduces the output ripple. Somewhat counter intuitively, the average output voltage decreases with the increase in decoupling capacitance. However, this additional average voltage due to ripple is anyway considered as loss in the literature as it doesn't provide any benefits to the supplied circuit.

By assuming a 50% duty cycle and that all the MOS switches have the same resistance, the output resistance of the voltage doubler can be calculated as

$$R_S = \frac{1}{2f_{sw}C} \frac{1 + e^{T_{sw}/2\tau}}{1 - e^{T_{sw}/2\tau}},\tag{3.57}$$

where $\tau = 2R_{on}C$ is the time constant of the doubler, accounting for the two switches in series with the capacitor, and C is the capacitance of the fly capacitors. The expression is valid for the entire range of switching frequencies from SSL to FSL. For the limit case, the SSL resistance does not depend on the switches and is equal to $1/(2f_{sw}C)$. Additional losses in the circuit are present in the form of $fC_{p,tot}V_{DD}^2$, where $C_{p,tot}$ is the sum of all parasitic capacitances, coming from the fly capacitors and gate capacitances of the switches. Assuming a constant current load, the efficiency is given by

$$\eta = \frac{1}{1 + R_S \frac{I_o}{V_o} + \frac{f_{sw} C_{p,tot} V_{DD}^2}{V_o I_o}}. \tag{3.58}$$

The output voltage can be calculated as $V_o = 2V_{DD} - R_S I_o$, allowing the load current that maximizes efficiency to be found. From the design perspective, for a given load current and the switching frequency (that might be determined by the system parameters) it is possible to find the capacitor size that results in a maximum efficiency. The switches should be sufficiently large to provide low on-resistance, but not too large to avoid excess switching losses due to their parasitics. With careful optimization it is possible to achieve efficiency in the order of 90%.

The shown voltage doubler can also be used as a voltage inverter, which is simply achieved by grounding the output of the doubler and using the source voltage of the NMOS switches as the output. This is a relatively efficient way to generate $-V_{DD}$. Negative voltage can be used to bias the bulk terminal of NMOS devices to reduce the leakage of circuits in the off state. The voltage doubler can be used to bias the N-well of the PMOS devices for the same reason. In fact, because the current into the bulk terminal is very small, a slow switching frequency can be used, making it possible to implement an SC converter that consumes nanoamps. For blocks with high leakage current, this approach could prove to be more beneficial than simply grounding the bulk and tying the N-well to V_{DD}.

Auxiliary Circuits

The core circuits of nearly every power management system are LDOs and DC–DC converters. However, we shouldn't forget that there a few other blocks that are necessary ingredients in every chip. These are the blocks that provide reference voltages and bias currents, as well as basic control signals, and ensure that the system powers-on in a correct way. In that sense these auxiliary circuits, as we will refer to them here, are essential as any small mistake might lead to the failure of the entire system. It might seem at first that this is a trivial, easy to implement part of the system, that is not crucial for achieving state-of-the-art performance but it plays an important role in the low-power sleep mode. Some of these auxiliary circuits have to remain on all the time as they provide the basic functions for the SoC and control the wake-up sequence. Knowing that the battery powered systems spend a lot of time sleeping, the consumption of the these auxiliary blocks will be a considerable factor in the overall system consumption, which is why their design should be handled with care.

4.1 System Overview

In a typical PMU, the DC–DC converter and the LDOs require different input signals that are generated by the auxiliary blocks. The first very obvious input is the reference voltage, needed to define their output voltage. An external circuit can be used for the purpose, but often an on-chip bandgap reference is used. The second important input is the bias current. It was not stated explicitly, but nearly every block that has been mentioned so far requires a bias current, an LDO error amplifier, a DC–DC converter compensator and comparators, to name a few. One convenient way to do this is to use one or a few current generators, and then to mirror this current and route it to the blocks that need it. Alternatively, it is possible to use the bias voltage and to route it to different blocks, but this approach should be avoided for two main reasons. The first is the voltage drop over the ground or supply lines that might result in a different voltage on the reference side and the side of the circuit using it. The second is the high-impedance of voltage nodes, which makes them susceptible to noise coupling.

The PWM controller of the DC–DC converter also requires a clock signal. The clock signal can be generated using a phase-locked loop (PLL) locked to the quartz oscillator. The quartz oscillator is present in nearly every SoC; it is used to generate the reference clock that is used for timekeeping and to derive other clocks needed by the SoC. Often, a full PLL is an overkill for a DC–DC converter clock, as it doesn't require high precision, and a simple RC oscillator might be sufficient. Because on-chip passive components can vary quite a bit from die to die, it is good practice to calibrate the frequency of these RC oscillators using the reference clock. Finally, the PMU will be controlled by digital signals. At a very minimum, nearly all circuits need an "enable" bit, used to turn them on or off. In the PMU context, there should be a signal indicating that the reference voltage and currents, as well as the clock, are stable and that the DC–DC converter can be turned on. Aside from the enable signal, it is often desirable to make some parameters programmable, allowing the circuits to be tested, and to tune their performance using software after production. For example, in a buck converter you could implement a programmable slope of the sawtooth waveform of the PWM modulator, allowing the loop gain to be tuned. Another option could be the gain of the current sensor, or even the passive components in the compensator, which could then be adjusted for different external components, making the design more versatile. As a general approach to design, the programmability of analog circuits allows for calibration via a digital controller or software, which provides the capability to compensate for mismatch, process and temperature variations. This approach is usually more power and area efficient than designing a fixed analog block that guarantees performance in all cases.

A simplified block diagram of an example PMU with the auxiliary blocks is shown in Figure 4.1. In this system multiple LDOs are used after the DC–DC converter. These LDOs isolate different parts of the system and act as supply switches that cut power to the unused blocks in order to minimize leakage in the off state. Leakage, especially in the case of digital circuits, can be a significant contributor to sleep-mode consumption in deep sub-nanometer

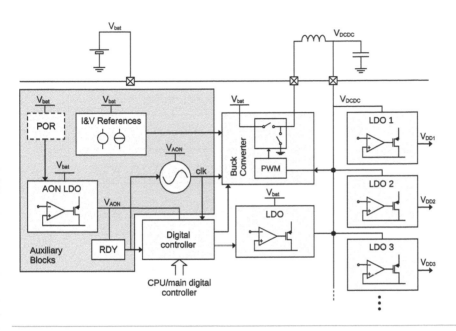

Figure 4.1: Block diagram of a power management system.

technology nodes, and the best way to reduce it is to disconnect the supply. A separate LDO is placed in parallel with the DC–DC converter. Note that it shares the IO pad and the external capacitor with the DC–DC converter. This is a common practice as the same chip may be used in different systems. The LDO might be more efficient if a battery with a sufficiently low supply voltage is used. Alternatively, you could imagine a scenario where highly sensitive circuits are operating and the LDO is used to provide the supply voltage to eliminate the ripple coming from the DC–DC converter.

As seen from the Figure 4.1, all the auxiliary blocks needed by the DC–DC converter must be powered either directly from the battery voltage, or from an auxiliary power source. Here the auxiliary power source is an always-on (AON) LDO that provides the regulated voltage to some of the blocks. As these blocks consume a relatively low amount of current the impact on the overall efficiency should be negligible. Use of an LDO to provide a regulated supply to auxiliary blocks is usually a good idea, as the battery voltage might vary quite a bit and make the design more complicated. The only block that absolutely must be powered from the battery at start-up is the voltage reference needed by the AON LDO.

Figure 4.2: Start-up sequence waveforms.

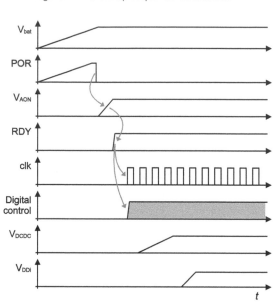

In general, for every PMU there is a sequence in which the blocks need to be turned on to assure correct operation. An example of such a sequence, valid for the presented system, is shown in Figure 4.2. As a general approach, once the battery voltage is established, the blocks will turn on one by one until the core supply finally settles at a correct value. At that point everything is ready for the CPU to start and take control of all the sub-systems. Often the first component to start in a system is the "power-on reset" signal (POR), which is high, meaning that it follows the rising slope of the battery voltage, until the supply reaches the minimum required value, and then drops to zero. The POR signal is typically used to keep the digital circuits in the reset state until the voltage is above a safe value, forcing the system into a known state upon powering up. Without the POR signal, register outputs would start from a random value, causing unpredictable behavior. Implicitly, a POR circuit requires a voltage reference to have something to compare the supply voltage to. It can be implemented as a standalone block, or integrated within an existing reference circuit. In the shown example the assumption is that the POR indicates that the references are stable and that the AON LDO can be turned on. The LDO needs a POR circuit of its own. Equivalently a "ready" (RDY) signal can be used, which is simply an inverted POR signal. Since the LDO start-up time is deterministic, it is possible to implement the RDY signal using a delay element, meaning that it would be a delayed, inverted POR signal, which could prove to be simpler than monitoring the LDO output voltage. We don't have this freedom with the external supply as it is beyond our control as IC designers. Here, the RDY signal starts the oscillator and the digital controller. It is a good practice to use the regulated supply for the these two blocks given that their performance are dependent on the supply voltage. In most cases the oscillator frequency varies at start-up and it is good to wait a few cycles for the frequency to stabilize before enabling the output clock. A counter that counts the required number of cycles can be used for this purpose. Once the clock signal is present, all conditions are met for the DC–DC converter to start, and the digital controller can initiate its own start-up sequence. When V_{DCDC} is ready, the LDOs that come after are powered and the core of the system can finally proceed with its intended operation.

The digital controller mentioned here controls only the PMU. As indicated in Figure 4.1 this controller is in fact configured by the CPU or the main digital controller of the chip, which is obviously switched off when the system is starting. A fixed default configuration must be used until the CPU is operational. To achieve this, "isolation" cells are placed at the controller inputs. These cells maintain a fixed output until the CPU is enabled, subsequently switching to the CPU defined value. Conceptually this is similar to placing multiplexers that switch between a fixed configuration and a CPU configuration. The only purpose of the fixed configuration is to power up V_{DCDC}. Since this is done once, when the battery is connected to the chip, the efficiency is not so much in focus as reliability is. This is the part of chip that must not fail. Once the core supply is established, more sophisticated blocks can be used to provide some of the auxiliary functions. For example, a relatively imprecise voltage reference can be used at start up, allowing to achieve V_{DCDC} that is just precise enough to power the core circuits. Later, an accurate bandgap reference could replace it, allowing to precisely regulate the DC–DC converter output.

The overview and the role of auxiliary blocks were presented in this section. In the following pages, we will focus on some common and practical ways to implement these auxiliary circuits and explain the fundamental concepts behind them.

4.2 Voltage and Current References

Voltage and current references are needed by almost every analog circuit out there, with specifications varying from one application to another. An example of a circuit that requires high precision, low-noise reference is an analog to digital converter (ADC). As for the power management, the requirements are often not that strict. Most analog circuits are designed to tolerate a $\pm 10\%$ supply variation. Accounting for the offset and mismatch inside the PMU, not more than half of it should be allocated for the voltage reference variation. The issue with standard CMOS technologies, especially the deep sub-nanometer nodes, is the process variation. The parameters of MOS transistors, resistors and capacitors vary quite a bit, making it difficult to implement a reference circuit with good absolute accuracy. The MOS threshold voltage will typically vary by $\pm 5\%$ at a fixed temperature, and have a small negative temperature coefficient. There are different resistor types available, which differ in sheet resistance and tolerance, with the resistance variation often in the range of ± 10–20%. MOS capacitors have a similar tolerance, while the metal capacitors (metal-oxide-metal or MOM capacitors) usually fall in the range of ± 5–10%. On the other hand, relative variation of components can be made quite small. If appropriate matching techniques are used, it can be reduced below 0.5%.

One component that still has relatively good characteristics when it comes to parameter accuracy, is the bipolar junction transistor. The BJTs available in standard CMOS technologies, are significantly worse compared to those in dedicated BJT technologies, so using them in amplifiers wouldn't be very effective. However, in voltage and current references, they still outperform MOS transistors, and are indispensable if a precise voltage needs to be generated. The main problem with BJTs is that the base-emitter voltage V_{BE} (essentially a pn junction voltage) doesn't scale with the technology node like the MOS threshold or supply. The threshold voltage of MOS devices decreases as the transistor feature dimensions decrease. As a result the supply reduces as well and it becomes more and more difficult to integrate BJTs with their V_{BE} which is constant around 0.7 V.

Most analog blocks, such as OTAs or operational amplifiers require a bias current. Fortunately, in most cases, the bias current can vary quite a bit without affecting circuit performance, allowing for a pure CMOS implementation. A PTAT (proportional to absolute temperature) current generator is shown in Figure 4.3 [31], the core of the circuit consists of transistors M_1–M_4. The drain currents of transistors M_1 and M_2 are equal, and set by the PMOS current mirror M_3–M_4. If the voltage headroom permits, a cascoded current mirror should be used for better precision. If transistors M_1 and M_2 are biased in weak inversion, the output current is given by

$$I_{ON} = I_{OP} = \frac{U_T}{R} \ln M, \tag{4.1}$$

where $U_T = kT/q$ is the thermal voltage and is proportional to the absolute temperature. Parameter M is the ratio of transistors M_1 and M_2. The output current is process dependent through resistor R. The slope of the current, with

Figure 4.3: PTAT current reference with a start-up circuit.

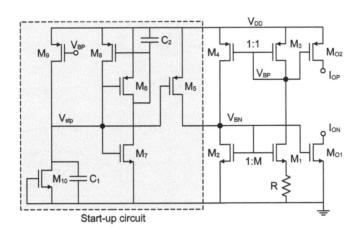

Start-up circuit

respect to temperature, can be tuned through M, R and the ratio of M_2 and the output NMOS M_{O1}. The PTAT circuit is also known as the constant G_m bias. The biased device should directly use the gate of the NMOS transistors. If the inversion factor of M_1 and M_2 corresponds to the inversion factor of the biased transistor, then the transconductance of the biased transistor equals $1/R$.

One problem that is found in all reference circuits is the presence of two stable points. The first one is the desired point, that provides the output PTAT current. In the second one, the currents in the two branches are 0, $V_{BN} = 0$ and $V_{BP} = V_{DD}$. To avoid getting stuck in the undesired point, a start-up circuit is needed. If the 0 state is detected, the circuit should pull the V_{BN} node up (or V_{BP} down), until a sufficient current is established in the two branches and the PTAT reference converges to the proper state. An example is shown in the left side of Figure 4.3. It is intended to provide proper start-up signals as the V_{DD} ramps up from 0. Initially, V_{BP} is at V_{DD}, M_9 is off and the node V_{stp} is pulled down through M_{10}. Transistor M_{10} is always off and presents a high resistance, it must be sized such that it is stronger than M_9 in the off state. This will allow the M_5 to pull V_{BN} up until the proper PTAT current is established. At this point, M_9 will start to conduct a scaled version of the PTAT current, pulling the node V_{stp} up. This switches the M_5 off and causes the inverter M_6–M_7 to flip. The latch M_6–M_8, will then maintain the node V_{stp} at a high value until V_{DD} is powered down, effectively disabling the start-up circuit once the desired state is reached. Capacitors C_1 and C_2 are used to prevent the latch from triggering at low V_{DD}, before the start-up circuit fulfills its role. The shown example illustrates the general concept for a PTAT reference, but it can be easily applied to a bandgap

Figure 4.4: Bandgap voltage reference.

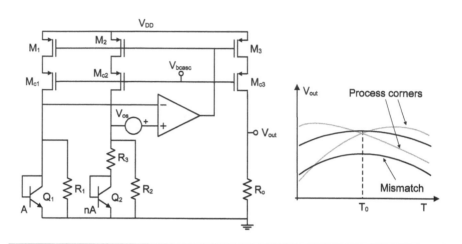

reference or modified to incorporate an external trigger (i.e. the reference is turned on using an enable bit, instead of powering on automatically when the supply is present).

The PTAT circuit from Figure 4.3 can be used as a part of a voltage reference [32], typically using the PTAT current to bias the MOS transistors and derive a reference voltage from the threshold voltage. Since both the resistor in the PTAT and MOS threshold voltage vary quite a bit, it is impossible to obtain an accurate voltage and temperature stability without calibration. This is where BJTs come into play, as they are much less sensitive to process variation and are able to provide an accuracy of 1–2% across PVT corners.

A low-voltage bandgap circuit is shown in Figure 4.4 [33]. In the shown implementation NPN transistors are used, but it is also possible to replace them with PNP transistors or diodes. As a general approach, all bandgap circuits use the same principle, where the reference is a weighted sum of two voltages, one with a positive and one with a negative temperature coefficient. The negative temperature coefficient comes from the V_{BE} of a bipolar transistor, and is typically close to -2 mV/K. The positive temperature coefficient comes from the difference of base-emitter voltages. Assuming that the equal current is imposed through the two diode connected BJTs with a size ratio of n (in the context of BJTs this is the ratio of emitter areas), the difference of base voltages is equal to $U_T \ln n$. The equal current is imposed using a current mirror. In Figure 4.4 a cascode current mirror is used in order to provide better current precision, however note that such an approach requires a higher supply voltage.

It is not necessary to use cascode transistors, but this approach results in better performance. The OTA is used to combine the V_{BE1} and the difference $V_{BE1} - V_{BE2}$ into a temperature independent voltage. In the shown example $R_1 = R_2$ and the output voltage is given by

$$V_{out} = \frac{R_o}{R_1} V_{BE1} + \frac{R_o}{R_3} U_T \ln n. \qquad (4.2)$$

The temperature coefficient is tuned using resistors R_1 and R_3 and the ratio n. Note that the output voltage does not depend on the absolute value of resistance, but only the ratio, making it process independent. The only absolute quantity is the V_{BE} which is fairly stable over process corners. This implementation is quite practical as any output voltage can be obtained. It is also possible to use it as a constant current source, although this current depends on the absolute value of resistors R_1 and R_3. The minimum supply needed for this circuit is the $V_{BE1} + V_{DS,sat}$, although a cascode current mirror is preferred, requiring at least $2V_{DS,sast}$ for the cascode transistors. When designing a bandgap circuit, the goal is typically to achieve a zero temperature coefficient at a nominal temperature T_0. Process variation will usually shift this point, and change the overall $V_{out}(T)$ curve, increasing the reference variation. A bandgap reference in a standard CMOS process should be able to maintain this variation within 1–2%. Often, a more important source of error is the mismatch, either coming from the current mirror or from the OTA offset. To illustrate the effect, it is possible to calculate the reference voltage, taking into account the OTA offset:

$$V_{out} = \frac{R_o}{R_1} \left(V_{BE1} + \frac{R_1}{R_3} U_T \ln n - \left(1 + \frac{R_1}{R_3} \right) V_{os} \right). \qquad (4.3)$$

The result is a shifted reference voltage and possibly modified curve, due to the temperature dependence of V_{os} itself. From the equation we can see that the way to minimize the impact due to the OTA offset is to minimize the R_1/R_3 ratio, but this alters the temperature coefficient. To compensate we then need to increase n, which finally leads to increased silicon area – a common end result when dealing with offset and mismatch. To deal with the OTA offset it also is possible to implement self-calibration, or to calibrate the circuit after production if high precision is needed. Additionally, a range of different BJT configurations can be found in the literature that provide different characteristics and allow to combat different non-ideal effects in the circuit. A minor detail, easily overlooked and worth mentioning here, is that the OTA of the bandgap needs a bias current, so a PTAT or another current generator is also required.

A similar BJT configuration can be used to implement a POR circuit. The OTA is replaced by a comparator and the current mirror is replaced by resistors used to sense the supply voltage. The voltages V_1 and V_2 rise at different

Figure 4.5: POR circuit based on the bandgap reference.

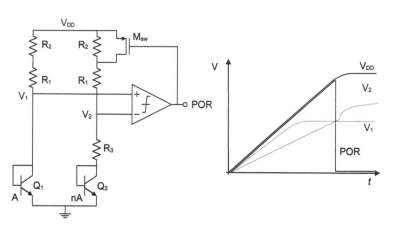

Figure 4.6: Supply detection circuit and RDY signal generation.

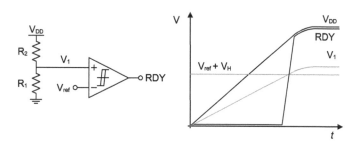

rates with the supply voltage. The voltage V_1 stabilizes at V_{BE1} first, whereas V_2 reaches the same point for a higher value of V_{DD}. This crossing point is the threshold that triggers the POR signal, as shown in Figure 4.5. Once the comparator output drops to 0, the transistor M_{sw} turns on and shorts the resistor R_2, resulting in a step up in V_2. This is equivalent to lowering the POR threshold, or implementing a hysteresis in the comparator, which improves the noise immunity of the circuit. It is a common technique used to prevent toggling of the comparator output due to supply noise. Aside from the hysteresis, it is a common practice to add a delay in the comparator output (e.g. using an RC filter) for the same reason.

If a reference is already active in the circuit, supply level can be detected using a comparator, as shown in Figure 4.6. Just like with the POR, a comparator

Figure 4.7: Schematic of a Schmitt trigger.

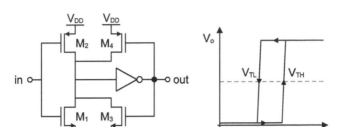

with hysteresis should be used. The resistive divider is used to tune the desired V_{DD} level that triggers the comparator. In this example a RDY signal is generated instead of a POR, with the only difference being the polarity of the signal. This circuit is primarily used to detect the minimum safe value at the output of an LDO or a DC–DC converter, signaling to the supplied circuits to turn on.

4.3 Comparators and Level-shifters

Comparators are used quite often in many different types of circuits. In the context of PMU they are used in the PWM modulator, for zero-crossing and high-current detection, as well as for voltage monitors in various places. In principle, any kind of OTA, or an operational amplifier can be used as a comparator. Design approach is different, as stability is not of concern in comparators, in fact positive feedback is commonly applied to increase gain or to add a hysteresis in the comparator input–output characteristic. One of the key parameters of a comparator is the delay that typically dictates the minimum bias current and power consumption. A symmetrical OTA [1] is a topology that is often used as a comparator as it already provides a rail-to-rail output. For other topologies, an inverter can be added to regenerate the output voltage level.

A Schmitt trigger is sometimes used at the OTA output, instead of a simple inverter, to avoid toggling of the output voltage when the two input voltages are close to each other. The circuit can also be used as an analog to a digital interface whenever a slow analog signal is driving a digital gate to improve noise immunity and prevent bouncing of the digital output. A schematic of a Schmitt trigger is shown in Figure 4.7. A positive feedback loop, that consists of the standard CMOS inverter and transistors M_3 and M_4, generates a difference

Figure 4.8: Schematic of a comparator with internal positive feedback.

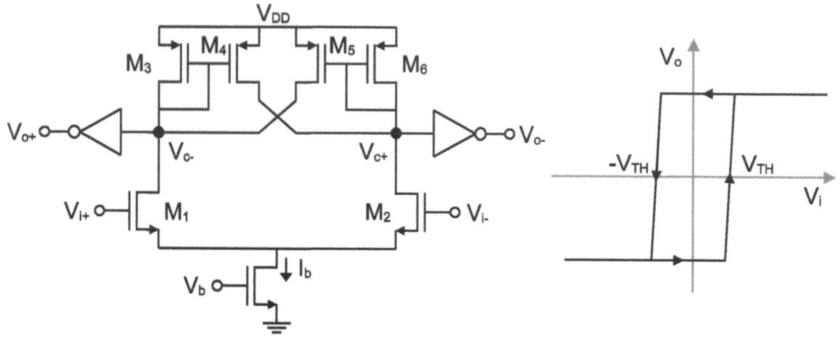

in the threshold voltages for the rising and falling edge of the input signal. The threshold of an inverter depends on the geometry of the NMOS and PMOS transistor. The intuitive way to look at the circuit is to see it as an inverter whose transistor geometry changes with the input state, resulting in the shift of the threshold voltage. When the input is at 0 so will be the output, meaning that M_4 is conducting together with M_2. For the Schmitt trigger to change state, the input voltage needs to be high enough that the NMOS M_1 overpowers the two PMOS transistors. Compared to a standard inverter consisting of transistors M_1 and M_2, the switching threshold V_{TH} is higher. The opposite is the case when the input voltage goes from V_{DD} to 0. Now the PMOS must overpower the two NMOS transistors, which means that the threshold V_{TL} is below the inverter threshold. Transistors M_3 and M_4 typically have a very small W/L ratio to assure that the input can flip the state of the Schmitt trigger.

Hysteresis in differential comparators can be implemented either by using an external resistive divider in a positive feedback configuration, or by embedding the positive feedback directly in the comparator, as shown in Figure 4.8. The positive feedback is generated using the cross-coupled transistors M_4 and M_5. The feedback factor is defined as $k = (\frac{W}{L})_4/(\frac{W}{L})_3 = (\frac{W}{L})_5/(\frac{W}{L})_6$. To understand the operating principle, you can assume that the input voltage $V_{in} = V_{i+} - V_{i-}$ is very low. The entire tail I_b current flows through M_2 and into a diode connected transistor M_6. Transistor M_5 is conducting and pulling the node V_{c-} to V_{DD}. As the input voltage increases, M_1 starts to conduct, however, its current is entirely absorbed by M_5 at first. At the threshold point the current of M_1 is equal to the current of M_5 (that is mirroring the current of M_6 at a ratio k). If the feedback factor $k > 0$ then this threshold is above 0 and hysteresis will occur. Further increase of current beyond the threshold point

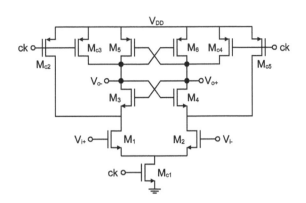

Figure 4.9: Clocked comparator – a StrongARM latch.

leads to conduction of M_3 and M_4, which consequently pull the node V_{c+} up to V_{DD} flipping the state of the comparator. The two thresholds are symmetric around 0 and the value is given by:

$$V_{TH} = \sqrt{\frac{2nI_b}{\mu C_{ox}W/L}} \left(\frac{\sqrt{k}-1}{\sqrt{k}+1} \right). \tag{4.4}$$

The threshold voltage depends on the transistor geometries, but also on the tail current I_b and the process parameters of the PMOS transistor. Although, in terms of area, this implementation is more efficient than using external resistors, the thresholds are dependent on process parameters and may vary, making it impractical for some applications. If $k = 0$ there will be no hysteresis. Note also that the cross coupled PMOS pair generates a negative conductance $-G_m$ in parallel with the positive conductance G_m of the diode connected devices. In theory, if the four PMOS transistors are equal, the impedance seen from the differential pair is infinite, resulting in an infinite gain of the comparator. This infinite gain is a consequence of the positive feedback. In practice these transistors will never be completely equal due to mismatch, and the gain will be finite. If $k < 1$ the circuit behaves as a standard gain stage which can be used as regular comparator.

So far we've discussed only continuous-time comparators, the comparators that change state as soon as the input voltage crosses the threshold. In some cases clocked comparators can be used instead. Their main advantage is that they consume no static current as they perform comparison only when triggered by a clock signal. Similarly to the CMOS circuits, the average consumption is proportional to the clock frequency. Typically, they can be useful inside

Figure 4.10: Schematic of a level-shifter.

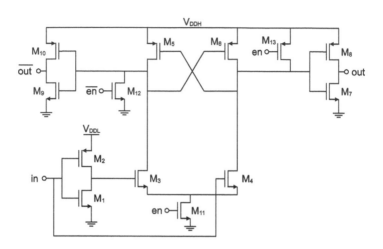

calibration loops or for monitoring slowly varying voltages (slow compared to the clock frequency). One example application is the regulation of the buck converter output voltage in a low power mode. Due to the large output capacitor and low load currents, it is sufficient to sample the output voltage at kHz frequencies, allowing for a very low-power implementation. The most commonly used clocked comparator is the StrongARM latch [22] shown in Figure 4.9, initially developed as a sense amplifier for memories. It consists of a differential pair and an output latch (M_3–M_6). In each clock cycle, the internal nodes are precharged to V_{DD} prior to the comparison via the switches M_{c2}–M_{c5}. At a rising edge of the clock, the difference in the input voltage creates an offset in the current of the differential pair that puts the output latch out of balance. The small disbalance is further amplified by the positive feedback, causing the circuit to lock in the desired state. At a falling edge of the clock the state of the comparator is erased and both outputs are pulled up. If the comparator state needs to be preserved throughout the entire clock cycle, it should be followed by an SR latch. When sizing the comparator, the input differential pair needs to be large enough to minimize offset, while all the other transistors can be close to the minimal size in order to reduce parasitic capacitances and increase the speed of the circuit.

The integrated PMU will inevitably have different blocks supplied by different voltages. For example, the controller of the buck converter my be supplied directly by the battery voltage, while the CPU supply comes from the converter and is lower than the battery voltage. For the CPU to talk to the controller, level-shifters must be placed between the two blocks to align the logic levels. A typical level-shifter that converts a signal from the low-supply domain V_{DDL} to the high supply domain V_{DDH} is shown in Figure 4.10. The circuit is somewhat similar to the comparators described so far. A pseudo-differential pair M_3–M_4 is used to flip the state of the latch M_5–M_6, that generates the output signal at the V_{DDH} level. Unlike the comparator that needs to detect a small input differential signal, here the input is a full swing digital signal, which makes the design simpler. The enable signal is often required in the level-shifter. In the case where the low supply is off, and the input signal undefined, the enable should be held at 0 which will maintain the output of the level-shifter at a predefined value. The feature is useful during start-up to maintain a fixed configuration of the DC–DC converter controller until the core supply is powered-on and the CPU ready to take over control (check Figure 4.1). In general, it is a good practice to put level-shifters between two supply domains even when there is no difference between the two voltages. This is simply done to avoid unpredictable behavior in the cases when one supply is off and another on.

4.4 On-chip Frequency References

Most systems on chip today use a quartz oscillator to generate the reference clock signal, which all the other clocks are derived from using, for example, a PLL. One of these clocks can be used for the PWM controller of the DC–DC converter. However, if the quartz oscillator itself is powered from the converter, a different oscillator must be used, at least until the converter supply is stable. An integrated RC oscillator can be used for this purpose. Since the oscillation frequency depends on the integrated components, the oscillator frequency is likely to be dependent on the process variation. However, the DC–DC converters are not sensitive to frequency variation, aside from perhaps a slight change in efficiency, so there shouldn't be any concerns for the start-up functionality. Once the system is powered on, there might be limits coming from other parts of the system (e.g. operating close to the frequency of another block), so a more precise clock may be needed. For high precision, it is possible to use a PLL, but very often it will be sufficient to calibrate the existing RC oscillator using the quartz reference. For this reason the RC oscillators are almost always implemented with a digitally tunable resistor or capacitor. Most solutions found in the literature primarily aim to reduce the process, supply and temperature variation of the clock frequency. Although this isn't necessary for DC–DC converters it is generally preferable to have the clock frequency confined to some limits. For a decent implementation, the clock variation after the initial calibration will usually be well within 1% for the entire temperature range.

Figure 4.11: RC oscillator with threshold voltage compensation.

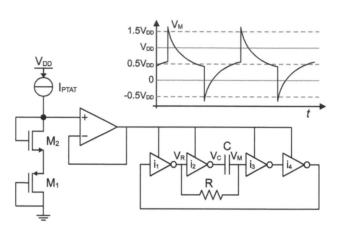

There are a huge number of ways to implement on-chip oscillators, with different implementations resulting in different characteristics. A class of oscillators commonly found in CMOS chips are the ring oscillators. They consist of an odd number of inverters tied in a ring. Owing to the advances in the CMOS technology, these oscillators can easily reach frequencies of several gigahertz, or can be used as power-efficient, low-frequency oscillators. The problem with such a structure is a strong dependence on the supply, temperature and process changes, and poor frequency stability. The two implementations shown here, use improved ring topologies, that address these issues.

The first oscillator, described in [34], is shown in Figure 4.11. The ring chain is cut in two with the capacitor C, it actually consists of two inverting elements, one from the resistor terminal V_R to the capacitor terminal V_C, and another from the common terminal V_M to V_R. The number of inverters is actually even, but note that one of them is bypassed by a resistor, resulting in an unstable circuit overall. In each period the node V_C toggles between 0 and V_{DD}, making the node V_M toggle between $-0.5V_{DD}$ and $1.5V_{DD}$, as shown in the figure. After the state flips, the capacitor is charged or discharged through the resistor until V_M reaches the threshold of inverter i_3, when the state changes again. Assuming the threshold is at $0.5V_{DD}$ and that the delay of inverters is negligible compared to the RC constant of the circuit, the period is equal to $RC \ln 9$. In practice, the inverter delay affects the oscillation stability and supply regulation is used to maintain it at a constant value. A flipped inverter biased from a PTAT current source provides a reference value for supply regulation. In a first order approximation, such bias should maintain the constant transconductance of

Figure 4.12: RC oscillator with an output phase combiner.

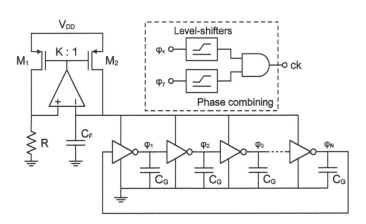

the inverter stages making the inverter delay independent of temperature and supply changes. The process variations (variations of R and C) must be calibrated separately.

The second oscillator is shown in Figure 4.12 [35]. It is similar to the first one in the sense that supply regulation is used to stabilize the oscillation frequency of a ring oscillator. An OTA in a feedback loop is used to set the the drain currents of transistors M_1 and M_2. The ratio of these currents is K, and can be chosen freely to optimize the size of the passive components. The current drawn by the ring oscillator is fNC_GV, where N is the number of inverters in the ring, C_G is the total gate capacitance in each node and V is the supply voltage of the ring oscillator (the negative input of the OTA). Knowing the ratio of currents in the two branches it is possible to calculate the oscillation frequency as

$$f = \frac{1}{KNRC_G}. \tag{4.5}$$

Again, the frequency is determined by an RC constant of the circuit. In the original form, this circuit was used as a temperature sensor [35]. A resistor with a strong temperature dependency would result in a frequency that changes with temperature. Here we have the opposite goal, to make the frequency temperature independent, which is achieved by choosing a resistor with a minimum temperature coefficient available in a given technology. The temperature dependency cannot be fully eliminated, but it can be made sufficiently small for the targeted application. Since the inverter outputs are below the supply V_{DD}, a level shifter should be used to provide a full swing clock signal that can drive standard digital gates.

Figure 4.13: Ring oscillator output phases and an example of phase combining.

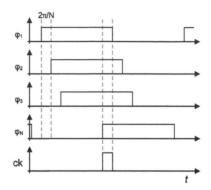

Another property of the ring oscillator is used here. For a ring oscillator with N stages, the output signal of each inverter will be shifted in phase by $2\pi/N$ compared to the previous inverter. The waveforms are illustrated in Figure 4.13. Different phases can then be combined using a simple AND gate (Figure 4.12) to obtain a desired duty cycle of the output clock signal. Such a circuit may be useful for a buck or boost converter with a constant on-time control. The larger the number of inverters in the chain the higher the resolution of clock pulse, making the circuit convenient for integration in a PMU.

References

[1] W. M. C. Sansen, *Analog Design Essentials*. Springer, 2006.

[2] B. Razavi, *Design of Analog CMOS Integrated Circuits*. McGraw Hill, 2005.

[3] C. C. Enz and E. A. Vittoz, *Charge-Based MOS Transistor Modeling: The EKV Model for Low-power and RF IC Design*. John Wiley & Sons, Ltd, 2006.

[4] T. Y. Chyan, H. Ramiah, S. F. W. M. Hatta, N. S. Lai, C.-C. Lim, Y. Chen, P.-I. Mak, and R. P. Martins, "Evaluation and perspective of analog low-dropout voltage regulators: A review," *IEEE Access*, vol. 10, pp. 114 469–114 489, 2022.

[5] Y. Lu, W.-H. Ki, and C. P. Yue, "17.11 a 0.65ns-response-time 3.01ps fom fully-integrated low-dropout regulator with full-spectrum power-supply-rejection for wideband communication systems," in *2014 IEEE International Solid-State Circuits Conference Digest of Technical Papers (ISSCC)*, 2014, pp. 306–307.

[6] J. Torres, M. El-Nozahi, A. Amer, S. Gopalraju, R. Abdullah, K. Entesari, and E. Sanchez-Sinencio, "Low drop-out voltage regulators: Capacitor-less architecture comparison," *IEEE Circuits and Systems Magazine*, vol. 14, no. 2, pp. 6–26, 2014.

[7] X. Liu, H. K. Krishnamurthy, T. Na, S. Weng, K. Z. Ahmed, K. Ravichandran, J. Tschanz, and V. De, "14.7 a modular hybrid ldo with fast load-transient response and programmable psrr in 14nm cmos featuring dynamic clamp tuning and time-constant compensation," in *2019 IEEE International Solid- State Circuits Conference - (ISSCC)*, 2019, pp. 234–236.

[8] V. Gupta, G. Rincon-Mora, and P. Raha, "Analysis and design of monolithic, high psr, linear regulators for soc applications," in *IEEE International SOC Conference, 2004. Proceedings*, 2004, pp. 311–315.

[9] X. L. Tan, S. S. Chong, P. K. Chan, and U. Dasgupta, "A LDO regulator with weighted current feedback technique for 0.47 nF–10 nF capacitive

load," *IEEE Journal of Solid-State Circuits*, vol. 49, no. 11, pp. 2658–2672, 2014.

[10]   P. Y. Or and K. N. Leung, "An output-capacitorless low-dropout regulator with direct voltage-spike detection," *IEEE Journal of Solid-State Circuits*, vol. 45, no. 2, pp. 458–466, 2010.

[11]   C.-J. Park, M. Onabajo, and J. Silva-Martinez, "External capacitor-less low drop-out regulator with 25 db superior power supply rejection in the 0.4–4 mhz range," *IEEE Journal of Solid-State Circuits*, vol. 49, no. 2, pp. 486–501, 2014.

[12]   G. den Besten and B. Nauta, "Embedded 5 v-to-3.3 v voltage regulator for supplying digital ic's in 3.3 v cmos technology," *IEEE Journal of Solid-State Circuits*, vol. 33, no. 7, pp. 956–962, 1998.

[13]   Y. Okuma, K. Ishida, Y. Ryu, X. Zhang, P.-H. Chen, K. Watanabe, M. Takamiya, and T. Sakurai, "0.5-v input digital ldo with 98.72.7-Âa quiescent current in 65nm cmos," in *IEEE Custom Integrated Circuits Conference 2010*, 2010, pp. 1–4.

[14]   Y.-J. Lee, W. Qu, S. Singh, D.-Y. Kim, K.-H. Kim, S.-H. Kim, J.-J. Park, and G.-H. Cho, "A 200-ma digital low drop-out regulator with coarse-fine dual loop in mobile application processor," *IEEE Journal of Solid-State Circuits*, vol. 52, no. 1, pp. 64–76, 2017.

[15]   R. W. Erickson and D. Maksimovic, *Fundamentals of Power Electronics*. Springer, 2001.

[16]   C. P. Basso, *Switch-Mode Power Supplies: Spice Simulations and Practical Design*. McGraw-Hill, 2001.

[17]   R. Ridley, "A new, continuous-time model for current-mode control (power convertors)," *IEEE Transactions on Power Electronics*, vol. 6, no. 2, pp. 271–280, 1991.

[18]   C. P. Basso, *Designing Control Loops for Linear and Switching Powr Supples: A Tutorial Guide*. Artech House, 2012.

[19]   C. Y. Leung, P. Mok, K. N. Leung, and M. Chan, "An integrated CMOS current-sensing circuit for low-voltage current-mode buck regulator," *IEEE Transactions on Circuits and Systems II: Express Briefs*, vol. 52, no. 7, pp. 394–397, 2005.

[20]   S. Stanzione, C. van Liempd, R. van Schaijk, Y. Naito, F. Yazicioglu, and C. Van Hoof, "A high voltage self-biased integrated DC-DC buck converter with fully analog MPPT algorithm for electrostatic energy harvesters," *IEEE Journal of Solid-State Circuits*, vol. 48, no. 12, pp. 3002–3010, 2013.

[21]   R. Redl and J. Sun, "Ripple-based control of switching regulators—an overview," *IEEE Transactions on Power Electronics*, vol. 24, no. 12, pp. 2669–2680, 2009.

[22]   B. Razavi, "The strongarm latch [a circuit for all seasons]," *IEEE Solid-State Circuits Magazine*, vol. 7, no. 2, pp. 12–17, 2015.

[23]   P.-H. Chen, C.-S. Wu, and K.-C. Lin, "A 50 nW-to-10 mW output power tri-mode digital buck converter with self-tracking zero current detection for photovoltaic energy harvesting," *IEEE Journal of Solid-State Circuits*, vol. 51, no. 2, pp. 523–532, 2016.

[24]   S. Bandyopadhyay, P. P. Mercier, A. C. Lysaght, K. M. Stankovic, and A. P. Chandrakasan, "A 1.1 nW energy-harvesting system with 544 pW quiescent power for next-generation implants," *IEEE Journal of Solid-State Circuits*, vol. 49, no. 12, pp. 2812–2824, 2014.

[25]   J. Katic, S. Rodriguez, and A. Rusu, "A high-efficiency energy harvesting interface for implanted biofuel cell and thermal harvesters," *IEEE Transactions on Power Electronics*, vol. 33, no. 5, pp. 4125–4134, 2017.

[26]   M. D. Seeman and S. R. Sanders, "Analysis and optimization of switched-capacitor dc–dc converters," *IEEE Transactions on Power Electronics*, vol. 23, no. 2, pp. 841–851, 2008.

[27]   T. Van Breussegem and M. Steyaert, *CMOS Integrated Capacitive DC-DC Converters*. Springer Science & Business Media, 2012.

[28]   M. S. Makowski and D. Maksimovic, "Performance limits of switched-capacitor dc-dc converters," in *Proceedings of PESC'95-Power Electronics Specialist Conference*, vol. 2. IEEE, 1995, pp. 1215–1221.

[29]   H.-P. Le, S. R. Sanders, and E. Alon, "Design techniques for fully integrated switched-capacitor dc-dc converters," *IEEE Journal of Solid-State Circuits*, vol. 46, no. 9, pp. 2120–2131, 2011.

[30]   P. Favrat, P. Deval, and M. Declercq, "A high-efficiency cmos voltage doubler," *IEEE Journal of Solid-State Circuits*, vol. 33, no. 3, pp. 410–416, 1998.

[31]   E. A. Vittoz and O. Neyroud, "A low-voltage cmos bandgap reference," *IEEE Journal of Solid-State Circuits*, vol. 14, no. 3, pp. 573–579, 1979.

[32]   K. N. Leung and P. K. Mok, "A cmos voltage reference based on weighted δv/sub gs/for cmos low-dropout linear regulators," *IEEE Journal of Solid-State Circuits*, vol. 38, no. 1, pp. 146–150, 2003.

[33]   H. Banba, H. Shiga, A. Umezawa, T. Miyaba, T. Tanzawa, S. Atsumi, and K. Sakui, "A cmos bandgap reference circuit with sub-1-v operation," *IEEE Journal of Solid-State Circuits*, vol. 34, no. 5, pp. 670–674, 1999.

[34]   D. Griffith, P. T. Røine, J. Murdock, and R. Smith, "17.8 a 190nw 33khz rc oscillator with±0.21% temperature stability and 4ppm long-term stability," in *2014 IEEE International Solid-State Circuits Conference Digest of Technical Papers (ISSCC)*. IEEE, 2014, pp. 300–301.

[35]   D. Ruffieux, F. Krummenacher, A. Pezous, and G. Spinola-Durante, "Silicon resonator based 3.2 μw real time clock with $\pm$10 ppm frequency

accuracy," *IEEE Journal of Solid-State Circuits*, vol. 45, no. 1, pp. 224–234, 2010.

[36]   K.-H. Chen, *Power Management Techniques for Integrated Circuit Design.* Wiley, 2016.

[37]   B. Razavi, "The low dropout regulator [a circuit for all seasons]," *IEEE Solid-State Circuits Magazine*, vol. 11, no. 2, pp. 8–13, 2019.

[38]   W. Xu, P. Upadhyaya, X. Wang, R. Tsang, and L. Lin, "5.10 a 1a ldo regulator driven by a 0.0013mm2 class-d controller," in *2017 IEEE International Solid-State Circuits Conference (ISSCC)*, 2017, pp. 104–105.

[39]   O. Trescases and Y. Wen, "A survey of light-load efficiency improvement techniques for low-power dc-dc converters," in *8th International Conference on Power Electronics-ECCE Asia.* IEEE, 2011, pp. 326–333.

[40]   A. Shrivastava, N. E. Roberts, O. U. Khan, D. D. Wentzloff, and B. H. Calhoun, "A 10 mV-input boost converter with inductor peak current control and zero detection for thermoelectric and solar energy harvesting with 220 mV cold-start and −14.5 dBm, 915 MHz RF kick-start," *IEEE Journal of Solid-State Circuits*, vol. 50, no. 8, pp. 1820–1832, 2015.

[41]   M. Liu, K. Pelzers, R. van Dommele, A. van Roermund, and P. Harpe, "A106nw 10 b 80 ks/s sar adc with duty-cycled reference generation in 65 nm cmos," *IEEE Journal of Solid-State Circuits*, vol. 51, no. 10, pp. 2435–2445, 2016.

Index